The End of Particle Age

Sandeep Chavan

Published by SANDEEP CHAVAN, 2025.

THE END OF PARTICLE AGE

First edition. November 24, 2025.

ISBN: 979-8232427283

Written by Sandeep Chavan.

Table of Contents

To every curious mind who questioned the contradictions—and waited patiently for a universe without fragments.

"The universe was never divided—only our understanding was."

Sandeep Chavan

Disclaimer & Statement of Originality

Disclaimer

This book presents an independent scientific and philosophical framework known as **Universal Energy Dynamics (UED)** and **Ripple Field Dynamics (RFD)** developed by the author. The interpretations, models, and explanations offered here do **not** represent the teachings, theories, or positions of any academic institution, scientific body, or mainstream physics tradition.

All comparisons with established theories — including quantum mechanics, relativity, particle physics, and cosmology — are made solely for the purpose of **critical analysis, reinterpretation, and conceptual clarity**. This book does not claim endorsement, validation, or alignment with existing scientific consensus.

Readers are encouraged to explore, question, and critically analyze the ideas presented. This work invites inquiry, not blind acceptance.

The technologies, future scenarios, and applications described in later chapters are **conceptual extrapolations** intended to illustrate the implications of a continuous-field worldview. They are not predictions or guarantees, but explorations rooted in the logic of UED/RFD.

This book is not a substitute for formal education in physics or engineering, nor does it offer commercial, medical, or legal advice of any kind.

Statement of Originality

All concepts, terminologies, models, frameworks, and interpretations presented in this book — including but not limited to:

- **Universal Energy Dynamics (UED)**
- **Ripple Field Dynamics (RFD)**
- **Structured Vacuum Energy (SVE)**
- **Ripple-knot, coherence architecture, gradient logic**
- **Mass as ripple rigidity; charge as curvature asymmetry**
- **Law of Consequences (LOC)**
- **Alignment propagation; tension redistribution**
- **Ripple logic machines; coherence computing**
- **Field engineering; SVE gradient propulsion**
- **Ripple-diagnostics; coherence medicine**
- **Continuous Universe Mindset**
- and related constructs —

are **original intellectual contributions** by **Sandeep J. Chavan.**

This work synthesizes independent research, field logic, philosophical reasoning, intuitive models, and reinterpretations of observed physical phenomena without relying on existing particle-based ontologies. Where mainstream experiments or equations are referenced, they are used strictly for reinterpretation within the author's own continuous-field framework.

The author asserts full creative and intellectual ownership over the structure, articulation, and theoretical formulation of UED, RFD, and LOC.

Any resemblance to existing models is coincidental or arises from the reinterpretation of universally known scientific phenomena.

Reproduction or use of these models, names, or frameworks without proper attribution is discouraged and may constitute a violation of intellectual integrity.

Author's Note

Why This Book Matters Now

Physics has always been a courageous field. It dares to ask questions that sound almost unreasonable: *What is the universe made of? Why does anything exist? What holds the world together?*

For more than a century, we believed the answer lay in tiny, indivisible building blocks—particles. Electrons, quarks, photons, neutrinos...a neat little zoo of objects doing a neat little dance. It was a comfortable story. Clear. Visual. Easy to teach. Easy to imagine.

But comfort is never the goal of science.

The more we observed, the more our "particles" behaved like something else. They scattered, dissolved, morphed, disappeared, reappeared, acted like waves, acted like nothing, and sometimes acted as if they only existed because we looked at them.

Something didn't add up.

Our equations worked. Our experiments worked. But our **story**—the conceptual narrative we used to explain the universe—was falling apart.

This book was born from that tension.

Not from rebellion. Not from mysticism. Not from the desire to destroy the beauty of modern physics.

But from a simple realization:

We are trying to describe a continuous universe with fragmented ideas.

We broke reality into "particles," "forces," "carriers," "objects," and "interactions" because we needed simplicity. But nature is not obligated to match our imagination. At some point, our explanations must grow up.

That time is now.

The discoveries of the last hundred years—from quantum fields to vacuum fluctuations to high-energy experiments—quietly whisper the same truth:

There is no particle. There is only the field.

Once you see this, everything becomes shockingly simple. The paradoxes soften. The mathematical patches disappear. Forces lose their mystery. Matter gains clarity. And nature regains its elegance.

This book matters now because:

- Physics is stuck between incompatible frameworks.
- Students are confused by what is real and what is metaphor.
- Engineers and technologists need a clearer conceptual foundation for the next century.
- Researchers are drowning in complexity that should not exist.
- And science lovers deserve a universe that actually makes sense.

"The End of Particle Age" is not an attack on physics. It is a celebration of physics.

It honors the observations. It respects the experiments. It preserves the data. But it changes the story.

We move from *objects* to *ripples*, from *forces* to *consequence*, from *fragmentation* to *continuity*.

And we do it in the simplest possible language—no mathematics, no jargon walls, no unnecessary complexity. Just a clear explanation of the universe as it actually behaves.

This book is written for:

- the scholar who senses something missing,
- the researcher tired of patchwork theories,

- the engineer who wants physics to be practical again,
- the student intimidated by contradictions,
- the teacher who wants clarity without compromise,
- and the everyday science lover who believes the universe should not require mental gymnastics to understand.

If you have ever looked at modern physics and thought, "Why does this feel more complicated than necessary?" you are holding the right book.

We are not closing a chapter of science. We are opening the next one.

Welcome to the end of the particle age—and the beginning of a universe that finally makes sense.

Preface

For more than a century, physics has lived inside a story it created for itself— a story of particles, forces, probabilities, and paradoxes. It was a brilliant story, ingenious in its mathematics, heroic in its experiments, revolutionary for its time.

But it was still a story.

The deeper we looked into the universe, the more contradictions we had to invent to keep that story alive: wave–particle duality, virtual particles, collapse, renormalization, dark matter, dark energy, singularities, vacuum fluctuations, force carriers, entanglement mysteries, and countless patches.

Every new discovery demanded another exception. Every contradiction demanded another layer of abstraction. Every anomaly demanded another particle to "explain" it.

Meanwhile, the universe continued to behave with absolute simplicity, absolute coherence, and absolute continuity.

This book begins from a different starting point.

Not on particles. Not on forces. Not on probability waves. Not on mathematical rescue missions.

But on the one truth every phenomenon keeps pointing to:

The universe is a single continuous field. There are no pieces. There are only ripples.

Once you accept this, every paradox in physics dissolves. Quantum confusion disappears. Relativity becomes intuitive. Cosmology stops needing invisible substances. Black holes lose their mythology.

Gravitation stops being "mysterious." Light stops being magical. Matter stops being fragmented.

Everything becomes the behavior of one underlying field— a Structured Vacuum Energy (SVE) field whose ripples, gradients, tensions, and coherences create everything we call "physical reality."

This is the foundation of **Universal Energy Dynamics (UED)** and **Ripple Field Dynamics (RFD)**—the frameworks that guide this book.

You are not about to enter a fantasy, nor a speculative tomorrow. You are entering a universe that has always existed this way— we simply did not look correctly.

If you are a student, this book will free you from the contradictions that make physics feel impossible. If you are a researcher, this book will offer clarity where mainstream models feel forced. If you are a scientist, this book will challenge the assumptions that have limited your field for decades. If you are simply curious, this book will show you that the universe is far simpler than textbooks make it seem.

You do not need advanced mathematics to understand a continuous field. You need only a fresh lens and the willingness to observe without inherited bias.

My promise is simple:

I will not tell you what to believe. I will show you what becomes obvious once particles are removed from the story.

We are standing at the end of the Particle Age and the beginning of something far more elegant— a scientific worldview without fragmentation.

Welcome to a universe that was never divided.

— **Sandeep J. Chavan**

How to Read This Book

A simple guide to help every reader—student, researcher, or enthusiast—navigate a paradigm-shifting text

This book is not written like a traditional physics textbook, nor like a popular-science story. It sits exactly between the two—clear enough for anyone curious, rigorous enough for those who think deeply.

To get the most out of it, read with the following approach:

1. Read in a continuous flow

The universe is continuous, and so is this book. Every chapter flows into the next. If you try to break it into isolated pieces, you will miss the clarity that emerges from the overall narrative.

Let the ideas wash over you as one fabric.

2. Do not compare every idea with textbook physics

Mainstream physics is built on particles, forces, and abstractions. This book operates in a field-first universe.

If you keep switching frameworks, you will create artificial conflict. Instead:

- read the argument fully
- hold the continuous worldview in mind
- compare later

Treat this as a fresh lens, not a supplement to old ones.

3. Allow intuition to guide understanding

UED and RFD rely on curvature, ripple behavior, gradients, coherence, and equilibrium. These are intuitive concepts—if you allow them to be.

Do not look for equations. Look for patterns. Let your mind feel the field instead of calculating it.

4. Use the reflective interludes

After each major part, reflective interludes help you reset the lens. They are not summaries—they are anchor points.

Read them slowly. They reposition your intuition before the next leap.

5. Visualize everything as ripple behavior

Whether it's electron stability, nuclear structure, gravity, light, quantum events, black holes, time, or technology—visualize them as ripples in a continuous field.

This mental model dissolves paradoxes instantly.

6. Do not rush—the mindset shift matters more than details

The greatest challenge is not content but unlearning. Particle-thinking has been drilled into us for a century. Give yourself time to transition into flow-based thinking.

This is a foundational shift, not a quick read.

7. Use the Appendices as a toolkit

The appendices are not optional. They provide:

- vocabulary
- diagrams
- reinterpretation guides
- technology implications

- AI discussion prompts
- experiment reanalysis tools

Use them whenever you feel stuck or curious. They support both beginners and advanced readers.

8. Keep a second mind open—the mind of a child

Children grasp ripples, waves, gradients, and flows effortlessly. Adults complicate them.

Approach each concept with curiosity, not caution. Let simplicity surprise you.

9. Read slowly, but think boldly

This book does not ask you to memorize. It asks you to question.

You do not have to agree with everything. But you should examine everything—especially the ideas you have always accepted without proof.

10. Remember the central truth

The universe was never divided—only our vocabulary was.

Every chapter returns you to this point. Use it as your compass. If an idea feels confusing, check whether you are thinking in fragments. Shift back to continuity. Confusion will dissolve.

Final Suggestion

Read this book not to understand physics, but to understand the universe you already live in.

Because once you see reality through the continuous field, you will never again believe in particles, forces, or paradoxes. You will see coherence everywhere.

And that is the beginning of the Field Age.

Introduction

Welcome to the Last Era of Particles

For more than a hundred years, the idea of particles has shaped everything we teach, everything we research, and everything we imagine about the physical world. From textbooks to laboratories, from classrooms to supercomputers, particles were treated as the fundamental pieces of nature's machinery—tiny, indestructible dots building everything from atoms to galaxies.

It was a beautiful idea.

It gave us intuition. It gave us language. It gave us a mental picture that made physics feel tangible.

But beauty is not the same as truth. And pictures are not reality.

The closer we looked, the more the particle story began to break down.

The "perfect dots" flickered in and out of existence. They behaved like waves, then immediately refused to be waves. They dissolved, merged, reappeared, and transformed into patterns that no "object" should ever be able to perform. They interacted without touching, communicated without traveling, and existed without existing—until the precise moment we measured them.

Particle physics became a museum of contradictions.

- Wave–particle duality.
- Spontaneous collapse.
- Force-carrier particles.

- Virtual particles.
- Quantum fields producing particles but also being particles.
- Mass coming from one field, charge from another, spin from nowhere.

We explained the illusions with more illusions. We patched paradoxes with more paradoxes. We built entire careers around saving an idea that nature quietly stopped supporting long ago.

This introduction marks the moment we step out of that era.

Not with aggression. Not with arrogance. But with clarity.

Why call it the "Last Era of Particles"?

Because physics is ready to move beyond its metaphors.

Modern experiments already show that the universe does not behave like a zoo of objects. The Standard Model itself admits—quietly, reluctantly—that fields are fundamental and particles are simply excitations. Yet we still teach and think in terms of little balls because they're convenient.

But convenience is not a scientific principle.

This book invites you to finally let go of the fragments.

To see the universe not as a collection of things but as a single, continuous field performing different patterns of behavior:

- **Stable ripples** (what we call matter)
- **Traveling alignments** (what we call light)
- **Locked distortions** (what we call charge and mass)
- **Gradient-driven motions** (what we call forces)
- **Coherence failures** (what we call decay)
- **Field tension rebalancing** (what we call interactions)

Once you make this shift, something surprising happens:

All the paradoxes vanish. All the contradictions soften. Physics becomes intuitive again. The universe becomes elegant again.

The field replaces the fragments.

This is not a revolution against physics; it is a revolution within physics.

Everything that works—data, experiments, equations—remains untouched. What changes is the *story* we tell about what we see.

Instead of saying:

- "A particle moved," we say
- "A ripple rebalanced."

Instead of saying:

- "A force acted," we say
- "The field resolved its tension."

Instead of saying:

- "Particles exchanged energy," we say
- "The field reconfigured its alignment."

This is the shift from discrete thinking to continuous thinking. From objects to patterns. From pushing and pulling to equilibrium and consequence. From myth to mechanism.

Why now? Why this book? And why does it matter?

Because science is entering a new phase— a phase that demands clarity, simplicity, and conceptual truth.

Scholars want a unified picture. Researchers want explanations without paradoxes. Students want to understand physics without mental knots. Engineers want models that actually predict and scale.

Science lovers want a universe that feels intuitive, not mystical. And physicists want a framework that matches what experiments already show.

"The End of Particle Age" is not just a title.

It is a statement of transition. From the particle-centric worldview that served us well— to the field-centric worldview that will serve us better.

This introduction is your invitation to step into that clarity.

A universe without fragments awaits you. A universe where nothing acts alone. A universe where everything is consequence. A universe where every phenomenon—tiny or cosmic—arises from the same living field.

Welcome to the last era of particles. And the first era of understanding the universe as it truly is.

Part I — The Fall Of The Particle Worldview

"We mistook ripples for rocks and shadows for objects. Only when the field speaks, does the universe make sense again."

Short Overview of Part I

For more than a century, physics built its imagination around particles—tiny, discrete building blocks believed to compose everything we see. This worldview shaped our models, our vocabulary, our experiments, and even our intuition. It worked well enough to explain early discoveries, but as science advanced, the particle story began to fracture under its own weight.

Part I dismantles that old picture gently, clearly, and scientifically.

We begin by tracing how the particle idea took root: bubble-chamber tracks misread as objects, classical intuition carried into quantum landscapes, and the psychological comfort of imagining the universe as a collection of tiny things. Then we introduce the universal field—continuous, dynamic, and indivisible—as the true substrate of reality. Finally, we reveal why particles were never fundamental in the first place: every so-called particle is simply a ripple, a distortion, or a stabilized pattern of the field.

This part prepares readers for the shift from fragmentation to continuity. It dissolves myths, clarifies contradictions, and shows how physics became trapped in metaphors that no longer serve us. By the end, the reader sees with unmistakable clarity that the particle worldview was never the destination—only a stepping stone toward a deeper, more unified understanding of nature.

Chapter 1 — A Century of Beautiful Mistakes

How Physics Fell in Love with Particles

Physics has always carried the comfort of tiny things. If you give a child a handful of Lego blocks, they will instantly imagine a universe made of pieces. It is natural. Humans like discrete things: beads, bricks, atoms, dots, pixels. We understand the world better when we can break it into parts. Physics grew up with the same instinct.

When the universe started giving us strange hints—sparks on photographic plates, bubbles dancing in chambers, bright flashes in detectors—we interpreted them with the oldest tool in our mind: *"If something moves, it must be a thing."* This single assumption shaped an entire century. A century of brilliant discoveries, yes. A century of breathtaking mathematics, absolutely. But also—a century of beautiful mistakes.

Not foolish mistakes. Not careless mistakes. But human ones. Intuitive ones. Understandable ones. The kind that illuminate just how extraordinary our minds are, even in error. This chapter is about those mistakes—and how they quietly built the particle age.

1.1 — When Physics First Saw the Invisible

Imagine being a physicist in the early 1900s. Everything feels fresh. Electricity is booming. X-rays are magical. Atoms are mysterious shadows. Theories are being written faster than textbooks can contain them. Then suddenly, in cloud chambers and spark chambers, you begin to see tracks—tiny scratches across photographic plates, delicate streaks cutting through vapor trails. They look like footprints of invisible travelers.

What else could they be? "Particles," someone said. Small things. Little bullets passing through matter. Like raindrops leaving trails in fog. And the world agreed. Not because the evidence demanded it, but because the mind recognized the pattern. Humans are pattern-seeking creatures. We have always believed that if you see a path, something walked on it.

And so the idea of particles—tiny units of existence—became irresistible. Particles were easy to imagine. Easy to teach. Easy to draw on the board. Meanwhile, nature quietly smiled and said nothing. Because what created the tracks might not have been tiny "things" at all. It might have been something far more continuous, far more fluid, far more elegant.

But once a story becomes familiar, we rarely question it. Physics fell in love with particles the same way humanity once fell in love with the idea that Earth was the center of everything. Not because the story was true—but because it felt right.

1.2 — The Birth of the Particle Zoo

The 20th century became a treasure hunt. Every new flash of energy in a detector was proclaimed a discovery. Every curve on a graph hinted at another resident of the "particle zoo." Physicists celebrated each particle as if discovering a new species: Muon, Pion, Kaon, Tau, Lambda, Omega, Charm, Strange, Bottom, Top.

It was thrilling. It was heroic. It was chaotic. By the 1960s, there were so many "fundamental particles" that even Richard Feynman joked, *"If I could remember all their names, I'd be a botanist."* This should have been a warning. When physics begins to look like taxonomy, something is wrong.

But the prestige of discovery was intoxicating. The narrative was too powerful:

- New particle = New physics
- New symmetry = New theory
- New track = New identity
- New peak = New object

Meanwhile, the experiments weren't actually showing "particles." They were showing ripple signatures, resonance peaks, field interactions, temporary distortions, and energy realignments. Yet every time a phenomenon appeared, we forced it into the same old box: *It must be a particle.*

The particle worldview was becoming a habit—a deeply rooted psychological reflex. And habits are hard to break, especially in science, especially when they seem to work. But cracks were forming. Those cracks would change everything.

1.3 — The Universe Starts Misbehaving

By mid-century, experiments began revealing behavior no particle should ever display.

The double-slit experiments showed light acting like a wave, then instantly acting like particles if we merely "looked." The photoelectric effect revealed light behaving as if delivered in packets—yet the packets carried no size or dimension. Electron interference demonstrated electrons "interfering with themselves," as if each one explored every path simultaneously. Entanglement showed two "particles" behaving like one even when separated by kilometers. Virtual particles appeared only in equations, never in reality, yet were treated as physical messengers of force. Decay events revealed objects popping into existence and vanishing without reason, as if nothing about them was truly "solid."

These were not the behaviors of little things. They were the behaviors of patterns, ripples, coherences, field distortions, and tension resolutions. They were the behaviors of something continuous—not discrete.

But instead of abandoning the particle picture, science patched it. Wave–particle duality. Probability clouds. Quantum foam. Virtual photons. Force carriers. Hidden variables. Renormalization. Patch after patch—beautiful, impressive, mathematically brilliant patches—but patches nonetheless.

The universe had stopped behaving like particles, but physics would not let go of the story. We were trying to fix the map instead of looking at the territory.

1.4 — Why We Held On So Long

If particles were the wrong idea, why did physics hold on to them for a century?

Psychological comfort. Tiny objects are easy to imagine. Fields and ripples? Much harder. We like to picture the universe as a machine made of parts. It matches our intuition. It feels mechanical and safe.

Educational momentum. Every textbook, lecture, and curriculum was built on particles. It takes generations to rewrite that foundation.

Mathematical convenience. Equations for "point particles" were easier to produce, even if nature itself was not made of points.

Prestige of discovery. Declaring a new particle brought fame, funding, and recognition. Declaring "just another ripple pattern" did not.

Fear of losing structure. Abandoning particles felt like abandoning physics itself. It felt dangerous, like losing the ground beneath our feet.

And yet—the one thing physics forgot to ask was: *Does nature care about what we prefer?* The field does not adjust itself for our comfort.

Reality does not reorganize itself to match our metaphors. The universe does not follow our imagination—it expects us to evolve ours.

Particles lasted as long as they did not because they were true, but because they were useful while physics was young. But science is grown up now. And grown-up science must choose truth over comfort.

The next step is not to destroy the particle idea, but to replace it with something far more coherent. A worldview where nothing is fragmented. Where "objects" are temporary ripples. Where "forces" are field tensions. Where "mass" is locked curvature. Where "light" is alignment. Where the universe is not built from things—but from one continuous field performing many behaviors.

In the coming chapter, we step into that new foundation—the universal field that quietly supported all of physics all along, waiting for us to notice.

Welcome to the next chapter: "The Universe as One Field."

Chapter 2 — The Universe as One Field

The SVE foundation, ripple logic, and unified ontology

Imagine waking up one morning and discovering that every wall in your house was just an illusion created by clever lighting—the rooms were never separate; the boundaries only looked real. That is exactly what happens when physics finally sees the universe as one continuous field. Everything we thought was separate—particles, forces, masses, charges, interactions—dissolves like walls in the morning light. Not because they were wrong, but because they were never fundamental.

Beneath every interaction, beneath every experiment, beneath every equation, there is something deeper, quieter, more elegant: a single field. Structured. Alive with ripples. Unified across all scales. The true substrate of reality. The moment you glimpse it, the familiar world softens and rearranges itself—not into chaos, but into clarity. The map and the territory finally begin to agree.

This chapter unfolds that field gently, clearly, and continuously—without mathematics, without jargon barriers, and without breaking the flow. We move, step by step, from the world we know into the field-based universe that was always there, waiting for us to notice. And as the illusions fade, what remains is simple, resonant, and alive: one field, countless behaviors, and a universe that finally makes sense.

2.1 — The Field Beneath Everything

Modern physics has hinted for decades that the universe is not built from objects. And yet we held onto objects because they made our imagination feel safe. But when we look closely—really closely—something remarkable happens.

The "empty space" we thought was nothing turns out to be full. Structured. Active. Responsive. Deterministic in its architecture, yet free in its rippling motion.

This universal medium—call it vacuum, call it background, call it fabric—but in this book, we call it the **SVE: Structured Vacuum Energy.**

It is not mystical. Not speculative. Not hypothetical. It is the simplest interpretation of the most powerful fact in physics: even where there is "nothing," something persists.

The SVE is always present, always stable, always supportive—the foundation upon which every dynamic phenomenon occurs. It is continuous like water, stable like still air, responsive like a stretched membrane, and profound like the ocean that carries every wave.

The SVE is not a thing in space—it is space. Once this is understood, everything that once seemed fragmented becomes unified.

Matter? A pattern within the field. Light? A disturbance passing through the field. Gravity? A gradient within the field. Charge? An asymmetry in the field's orientation. Force? A consequence of the field resolving tension.

There is no outside of the field. There is no behind it. There is no deeper layer. The universe is one continuous field playing out its dynamics. Everything else is commentary.

2.2 — Ripple Logic: The Language of Reality

If the universe is one field, then its behavior must arise from patterns and motions within that field. This brings us to ripple logic—the natural behavior of disturbances in a continuous medium.

In everyday life, we see ripples in water. We hear ripples as sound. We feel ripples as vibrations. But the ripples of the SVE are not like water waves. They are far more fundamental. Some ripples propagate, as light and radiation. Some ripples lock, becoming matter. Some ripples dissolve, fading as unstable modes. Some ripples compress, manifesting as mass. Some ripples shift phase, appearing as charge. Some ripples align patterns, producing interactions.

Suddenly, everything the particle age struggled with becomes intuitive. A ripple is not an object. It has no edges, no borders, no rigid identity. It is a behavior. What we once called "particles" were particular kinds of sustained ripples—just stable enough to appear as enduring structures. What we called "forces" were simply changes in ripple tension—the field adjusting itself toward equilibrium.

Ripple logic becomes the universal language of physics. A decay is the collapse of ripple coherence. A collision is the superposition of ripple patterns. Mass is locked ripple density. Spin is the swirl topology of the ripple. Charge is orientation asymmetry. Light is alignment propagation.

No particle needs to "carry" anything. No messenger is required. No smallest object needs to exist. The sequence is simple and complete:

Field → Ripple → Behavior.

That is the ontology of reality.

2.3 — The SVE Ontology: One Universe, One Substance

Ontology asks the deepest question: *What is fundamentally real? What exists at the most basic level?*

For the particle worldview, the answer was fragmented—electrons, quarks, photons, neutrinos—each treated as a separate thing with its own identity. But for Unified Field Ontology (UFO—your future term), the answer becomes stunningly simple: **only the field exists. Everything else is its expression.**

This single statement resolves a hundred years of confusion.

How do particles pop into existence? Because ripples can spontaneously form in a continuous field under the right conditions. How do they vanish? Because ripples can unwind into the field just as easily. Why do particles exhibit wave–particle duality? Because ripples are not objects; they simply behave differently depending on boundary conditions. Why does entanglement exist? Because a continuous field cannot separate itself—information is shared by nature. Why does gravity act without touching? Because a field gradient affects everything within its domain. Why does the speed of light stay constant? Because the field sets the propagation conditions, not the "packet." Why do unstable particles decay so quickly? Because their ripple patterns fail the coherence test. Why do heavy particles require enormous energy to create? Because compressed ripple modes demand extremely high field distortion.

All these questions collapse into one elegant ontology: **the field does everything, particles do nothing.**

This is not philosophy. It is the only interpretation that aligns with every known experiment—without paradox, without conflict, without intellectual acrobatics. Once we adopt this ontology, the universe becomes not only understandable, but predictable for future science.

2.4 — Visualizing the Field: A Universe Without Edges

A field-based universe is hard to imagine at first, because the mind has been shaped by objects. Yet there are simple ways to visualize it.

Imagine a calm lake. Throw a pebble—ripples spread. Create many disturbances—patterns emerge. Some patterns collide, merge, swirl, stabilize. But there is never a new "object" in the water. Only behaviors. Matter is the stable swirl. Light is the traveling ripple. Gravity is the slope of water created by disturbances. Charge is the orientation of the swirl. Magnetic effects are the swirl's internal rotation. Decay is when a swirl cannot hold shape.

This analogy is simple but powerful. It dissolves the illusion of particles without losing the clarity of explanation.

Or imagine stretched fabric. Pinch it—ripples form. Pull it—tension aligns. Twist it—patterns lock. Release it—everything relaxes. That is the entire universe. Except the real field is far more precise, far more stable, and far more dynamic.

If the universe is one field, then the next natural question is: why did we believe in particles at all? Now that the field is understood, we need to reinterpret what we thought we observed. Tracks, peaks, decays, scattering events—none of these were particles. They were ripple signatures.

In the next chapter, we gently dismantle the century-old illusion of particles and replace it with a field-based view that is simpler, cleaner, and more elegant.

Chapter 3 awaits: "The Particle Illusion."

Chapter 3 — The Particle Illusion

Why We Never Actually Observed Particles

If you walk through a foggy street at night, the streetlamps look like floating spheres of light. They seem solid, round, almost tangible. But reach out to touch one, and you discover it was never a sphere at all—just light interacting with droplets of mist.

Physics has been walking through that fog for a century. We saw flashes, streaks, dots, spikes on graphs, sudden appearances, sudden disappearances—and assumed something tiny must be doing something solid. We named those illusions "particles."

But what if we only saw the fog? What if we never touched the real lamp? What if we confused a signature for a thing?

This chapter reveals an uncomfortable but liberating truth: we never actually observed particles. We only observed ripple effects inside a field. Everything we thought was a particle was simply a behavior—a ripple, a distortion, a local field event.

Once this is understood, the entire particle worldview dissolves gracefully, like mist in sunlight.

3.1 — Seeing is Not Observing: How Experiments Deceived Us

Physics experiments do not show us objects. They show us effects—changes within detectors, marks on plates, flashes in sensors.

Consider the famous bubble chamber. A charged disturbance moves through a superheated liquid and leaves a trail of bubbles. The human mind sees that trail and instantly imagines: *"Something moved through here."*

But what did we actually observe? A streak. A line of tiny bubbles. A pathway of disturbance. We never saw a "particle." We saw the field disturbing the medium.

Imagine dragging your finger through still water. You leave a trail. But the trail does not prove your finger is a particle. It only proves something caused a ripple.

In the same way, tracks are signatures, not entities. Hits are responses, not objects. Peaks in energy graphs are resonances, not species. Scattering patterns are interference, not collisions. Decay events are field reconfigurations, not objects splitting.

Modern detectors—CMS, ATLAS, LHCb, Belle—record digital patterns, not particles. We interpreted those patterns through the lens we inherited from classical mechanics. We did not observe particles. We inferred particles. And those inferences were based on a worldview the universe never confirmed.

3.2 — Light Without Particles: The First Clue We Ignored

One of the earliest signs that the particle picture was flawed came from the nature of light. Light produced interference like a wave. Light delivered energy like a particle. So physics invented the term *photon*—a compromise, a hybrid, a conceptual patch.

But a photon has no size, no boundary, no shape, no rest state, no internal structure. It appears only at the instant of interaction with matter. Never before. Never during propagation. A photon does not exist between emission and detection—only the field does.

What we interpreted as "photon arrival" was a ripple alignment event, the field handing off energy, a phase transition at the surface of the detector, a local consequence of global field propagation.

The photon was never a tiny bullet. It was the universe whispering in ripple-language, and we insisted it was speaking in discrete dots. The mistake was understandable. But it was still a mistake.

3.3 — The Strange Case of Particles That Behave Like Waves

If particles truly existed as tiny objects, they should behave predictably as objects do. But experiments kept telling a different story.

Electrons interfere with themselves. Fire one electron at a time through a double-slit, and the interference pattern emerges as if each electron passed through both slits. Objects do not do that. Patterns do.

Neutrinos oscillate identities in mid-flight. A neutrino starts as one "type," travels a few kilometers, and arrives as another. Objects do not morph like that. Fields do.

Virtual particles appear and vanish. No experiment directly detects them. They are mathematical corrections, not entities. Patterns do not need solidity. Particles supposedly do.

Quantum tunneling defies classical logic. How can a particle appear on the other side of a barrier without crossing the barrier? Simple: there was never a particle—only a ripple distribution that realigned.

Every act of quantum misbehavior becomes perfectly normal once you remove the object and keep only the ripple. The universe was not breaking rules. We were mislabeling the rules.

3.4 — High-Energy Collisions: Mistaken Identity at the LHC

When two protons collide at near-light speed, vast amounts of energy erupt into new patterns inside detectors. These patterns are interpreted as new particles, decays, annihilations, and exotic short-lived species.

But what is actually happening? Two tightly locked ripple architectures are smashed, their internal tensions unleashed. The field reacts by forming temporary distortions, creating resonance peaks, releasing alignment pulses, and snapping back into stable configurations.

These momentary distortions—lasting only 10^{-24} seconds—are then named: W boson, Z boson, Higgs boson, Tau lepton, Delta baryon. But calling these transient ripples "particles" is like calling a splash a new species of water. Nothing new was created. Only the field was disturbed.

Physicists know this, yet the particle narrative persists because it is easy to teach and easy to imagine. The detectors never see objects. They only see field events translated into digital blips. The universe never gave us particles. It gave us ripple signatures, and we interpreted them as tiny things.

3.5 — The Mind's Need for Objects: Why We Invented Particles

Even if nature did not give us particles, the human mind desperately wanted them. Why?

Objects are intuitive. Particles feel like miniature versions of stones and bullets. They match classical instinct.

Objects are easier to communicate. "Electron" is easier to say than "stable ripple-knot in the SVE field."

Objects allow simplification. A particle feels like a fundamental building block. A field feels too abstract.

Objects give narrative comfort. Humans love stories of interaction between actors. Particles provided characters.

Objects preserve identity. The idea that something has a "nature" is psychologically satisfying. Fields do not have natures—only behaviors.

The truth was harder to teach: reality is not made of things. Reality is made of consequences in a continuous field. We preferred a lie that felt like truth over a truth that felt too abstract to hold. And so the particle illusion survived.

3.6 — The Simpler Picture: Everything Is a Field Event

Once you let go of particles, the universe becomes astonishingly coherent.

- What we called "particles" are stable ripple distortions.
- What we called "forces" are field tension adjustments.
- What we called "mass" is locked ripple density.
- What we called "charge" is phase asymmetry in a swirl pattern.
- What we called "spin" is topology of ripple circulation.
- What we called "decay" is coherence failure in a ripple pattern.
- What we called "annihilation" is field relaxation into stable modes.
- What we called "vacuum" is the underlying SVE substrate.
- What we called "virtual particles" are mathematical echoes of ripple interactions.

Everything becomes continuous. Nothing needs discreteness. The universe stops being a zoo and starts being a sea.

If particles were illusions, then the next question becomes urgent: what then is matter? What keeps structures stable? Why do some ripples lock while others dissolve?

To answer that, we must look at the stable ripple-knots—the structures we once called electrons, protons, and neutrons. They are not particles. They are not objects. They are not fundamental bricks. They are stable behaviors of the field.

In the next chapter, we reveal the true nature of matter—not as pieces of the universe, but as long-lived ripples inside a single continuous field.

Chapter 4 awaits: "Electrons: The First Stable Ripple-Knot."

Interlude — When the Walls Begin to Disappear

A reflection at the end of Part I

There is a moment in every scientific journey when the old picture begins to fade, but the new one has not fully arrived. That is where we stand now—between two worlds, between two ways of seeing, between two stories we have told ourselves about the universe.

Part I was not about destroying physics. It was about removing the scaffolding that once helped us climb. The particle worldview was a brilliant bridge a conceptual stepping stone that carried us from classical solidity into the trembling landscape of quantum behavior. It gave us language. It gave us progress. It gave us confidence.

But now, standing at the other side, we can look back and see the cracks that were always there. Particles that acted like waves. Waves that behaved like things. Objects that dissolved when observed. Identities that changed mid-flight. Forces that acted without pushing. Vacuum that refused to be empty.

We were not wrong. We were simply looking through a keyhole, trying to describe a palace by the sliver of light that entered through it. Part I widened the keyhole.

We saw that the trail is not the traveler, the flash is not the thing, the peak is not the entity, the decay is not the breakup of a miniature object. We saw that experimental footprints do not guarantee the existence of little feet. And more importantly—we saw that the universe never truly promised us discreteness.

It only promised behavior, patterns, responses, ripples, tensions, alignments. We invented particles because the mind prefers pieces. Nature never confirmed our imagination. It simply allowed us to borrow the metaphor until we were ready for something clearer.

Now we are ready.

Part I asked us to gently loosen our grip on the particle worldview—not to reject it with anger, but to release it with gratitude. To say: *"Thank you for taking us this far. But you cannot take us where we need to go next."*

The next steps require a new lens—a lens wide enough to hold continuity, simple enough to avoid paradox, deep enough to be universal, and flexible enough to unify everything from electrons to galaxies. That lens is the field.

A continuous, structured, ever-present substrate in which every phenomenon—from stability to motion, from mass to light, from attraction to decay—is nothing more than a ripple performing the universe's choreography.

This interlude marks the transition. From fragments to flow. From objects to patterns. From observation to understanding. From noise to clarity.

You have stepped out of the Age of Particles. Ahead lies the continuous universe—more elegant, more coherent, more intuitive than anything the fragmented worldview ever allowed us to imagine.

As we enter Part II, we no longer ask: *"What is the particle made of?"* but instead: *"What does the field need to do for this structure to exist at all?"*

The universe is about to answer—not in pieces, but in ripples.

Part II — Rewriting Matter From The Field Up

"Matter is not a thing the universe builds—it is a rhythm the field chooses to sustain."

Short Overview of Part II

If Part I dissolved the illusion of particles, Part II reveals what truly stands in their place.

Here, we rebuild our understanding of matter from the ground up—not as a collection of tiny objects, but as a hierarchy of ripple behaviors within a single continuous field.

We begin with electrons, the simplest stable ripple-knot, and trace how their locked curvature gives rise to charge, mass, and structure. From there, we explore the composite architectures inside protons and neutrons, seeing how internal sub-ripples—misinterpreted as quarks—achieve nuclear stability. We then travel into the realm of pure propagation, where photons emerge not as particles, but as alignment waves carrying energy without ever condensing into "things."

Part II exposes the greatest conceptual misunderstanding of 20th-century physics: quarks and gluons were never particles at all, but internal distortions within larger ripple structures. Heavy modes—W, Z, Higgs, and the high-energy flickers observed in accelerators—are shown as temporary tension spikes of the field, unable to lock, appearing only for a heartbeat before dissolving back into equilibrium. And finally, neutrinos emerge as the field's lightest whisper—minimal distortions that travel immense distances while barely interacting.

This part is the heart of the book: a complete reinterpretation of all known "matter" as stable, semi-stable, or fleeting configurations of the SVE field.

No particles. No objects. No separate species.

Just one field performing many dances, each chapter revealing how nature builds complexity not from pieces, but from patterns.

When Part II concludes, you will see matter as the universe itself sees it—a spectrum of ripple-knot behaviors unified by the logic of the field.

Chapter 4 — Electrons

The First Stable Ripple-Knot

How locked curvature creates stability

If you shake a rope, the ripples travel and die out. If you fling water into the air, it scatters. If you strike a surface, the vibrations fade.

But imagine a ripple that does not fade. A ripple that curls into itself, holds its own shape, and becomes a kind of standing knot in the medium. Not because it is a separate object—but because the medium allows it to stay that way.

That is the electron.

Not a particle. Not a point. Not a tiny bead orbiting a nucleus. But the first stable ripple-knot the universe ever created—a perfectly balanced configuration of curvature inside the SVE field.

It is the simplest form of "matter," and the earliest clue that the universe does not build with bricks, but with patterns that refuse to collapse.

This chapter reveals how the electron achieves that stability, why it became the first long-lived ripple in cosmic history, and how this one knot of curvature holds the keys to chemistry, electricity, magnetism, and ultimately—life itself.

4.1 — The Electron Appears Before Atoms, Stars, or Light

To understand the electron, we must step back into the earliest moments of the universe—before stars lit, before atoms formed, before photons traveled freely. In the raw, expanding field, energy rippled in every direction. Most ripples dissolved instantly—their shapes too unstable, their patterns too jagged, their curvature unable to hold.

But a few ripples found a configuration that could lock. A swirl. A curvature that folded into itself, balanced by the SVE density around it. This was the electron: the universe's first successful attempt at sustained curvature.

Why did this specific configuration survive? Because the electron represents the simplest state where curvature is continuous, tension is balanced, energy is trapped but not exploding, the ripple does not collapse into stillness, and the surrounding field supports its geometry. Every other possible distortion failed these tests. The electron passed all of them.

It is the first creature of stability in a world that had only chaos. And because of this, the electron becomes the foundation upon which the universe later builds: atoms, molecules, solids, liquids, stars, electricity, magnetism—everything that lasts begins with a ripple that refused to disappear.

4.2 — Locked Curvature: The Secret Behind the Electron's Stability

Let us visualize the SVE field as a continuous medium—not a fluid, not a fabric, but a universal substrate. Now imagine a ripple forming in it. Most ripples spread outward and die. But the electron does something unusual: it curls back into itself. Not like a vortex that has a hole, but like a perfectly closed swirl of tension—a standing pattern of rotation where the curvature is so balanced that the ripple never unwinds.

This "locked curvature" is the electron's true identity.

The ripple wants to expand. The field wants to compress. Stability happens exactly where they cancel.

At that point:

- The ripple no longer spreads.
- The swirl no longer tightens.
- The energy no longer escapes.
- The pattern becomes self-sustaining.

This is mass. This is charge. This is stability.

Mass emerges because energy is trapped in a repeating curvature—like a drumhead vibrating within a fixed boundary. Charge emerges because the curvature has a directional twist—a field asymmetry that creates outward tension. Spin emerges because the topology of the swirl is helical—a structural property of how the ripple circulates within the field.

The electron does not *have* properties. The electron *is* those properties—each one arising naturally from the geometry of the knot. This is why its properties never change: you cannot alter the behavior of the knot without destroying its stability entirely.

The electron is a perfect example of nature's first mastery of equilibrium.

4.3 — Why the Electron Never Decays

Every other so-called "particle" seems to decay, transform, merge, split, or vanish. But the electron? It just stays.

Why? Because the electron's configuration is the simplest possible locked ripple that the universe can support. If you try to reduce it, the pattern collapses into pure SVE stillness. There is no simpler stable knot. If you try to excite it, a photon emerges from the disturbance, but the electron remains intact. If you try to smash it, you distort the field around it, yet the knot simply reforms itself—as long as the peak curvature remains.

In essence, the electron is the minimum unit of stable ripple curvature the field can maintain. That is why it does not decay. There is nowhere "lower" to go. Its stability is not a property—it is the geometry of survival in a continuous field.

That is also why the electron's mass, charge, and spin are universal constants. You cannot tweak the knot. You can only destroy it or let it be.

This explains why electrons across the entire universe behave identically, why atoms across galaxies follow the same chemistry, and why electricity functions consistently everywhere. When the geometry is fundamental, uniformity becomes unavoidable.

4.4 — Visualizing the Electron Without Particles

Let us visualize the electron in a way that anyone can grasp.

Picture a whirlpool in a perfectly calm lake. The water never drains. The swirl never grows. It simply sustains itself because the inflow and outflow are in perfect balance. But unlike a whirlpool, the electron has no hole, the swirl is closed, the curvature is locked, the pattern is three-dimensional, and the knot is stable indefinitely.

Or think of a guitar string plucked just right. Most vibrations fade. But strike the exact resonance, and the string vibrates in a standing pattern—stable, repeating, self-contained. The electron is the field's first resonance—the simplest standing pattern that does not fade.

Or imagine a tornado that never moves. A pattern of swirling energy held in place entirely by internal and external balance.

These analogies are imperfect, but they give a sense of why the electron persists without needing a tiny "particle" inside it. There is no "thing" in the electron. There is only a stable pattern. Matter begins as rhythm, not object.

4.5 — The Electron's Role in Building the Universe

Once the electron stabilizes, the universe finally has its first reliable building block. From that moment onward, everything begins to assemble.

Electrons orbit the earliest nuclei, giving atoms their identity. They weave chemical bonds, stitching molecules together. They carry electricity across vast networks, generate magnetism through their collective alignment, and define the structure of solids by locking into repeating patterns. They mediate interactions between molecules, produce light through subtle shifts in alignment, and ultimately enable life itself by enabling chemistry.

In field language, the story is even simpler. Atoms exist because electrons stabilize around nuclei. Molecules exist because electron ripples overlap coherently. Solids exist because electron knots align in ordered patterns. Conductivity exists because electrons unlock their positions and flow. Magnetism exists because countless electron swirls align in unison. Biology exists because electrons form complex, stable configurations that refuse to collapse.

All of this—atoms, molecules, stars, chemistry, life—emerges from one simple knot of curvature. The electron is the universe's first rhythm of stability, the swirl that endures. Everything else is assembly.

If electrons are the simplest stable knots, protons and neutrons are the next great leap— far more complex, far more energetic, with internal sub-ripples locked in a precise dance that mainstream physics misinterpreted as "quarks."

In the next chapter, we reveal how matter becomes dense, how internal curvature builds structure, and why the field, not particles, determines the behavior of atomic nuclei.

Chapter 5 awaits: "Protons & Neutrons: Composite Ripple Architectures."

Chapter 5 — Protons & Neutrons

Composite Ripple Architectures

Internal sub-ripples, confinement, and nuclear coherence

If the electron is the universe's first successful ripple-knot, the proton is the universe attempting something far more ambitious.

Imagine trying to tie not one, but three swirling knots of tension inside the same field—each one tugging at the others, each one trying to escape, and yet somehow forming the most stable and essential structure in the visible universe.

This is the proton: nature's masterpiece of field engineering.

Not a spherical bead. Not three quarks trapped inside a prison. Not a tiny bag of force-carrying gluons. But a dense, composite ripple architecture—a multi-layered swirl pattern where internal distortions are locked into stable harmony.

The neutron is its close cousin: almost the same architecture, but with a subtle imbalance that makes it stable only when surrounded by the right field conditions.

In this chapter, we step inside these remarkable ripple structures—to see how the universe builds density, how internal sub-ripples create what we mistook for quarks, and how nuclear coherence emerges from field logic, not particle confinement.

5.1 — The Proton: The Field's Densest Stable Knot

After electrons stabilized in the early universe, the field attempted more complex configurations. Most collapsed instantly—too unstable, too asymmetric, too energetic. But one configuration held: the proton.

If the electron is a simple swirl, the proton is a tri-ripple architecture—a locked combination of three internal swirls, each one creating and responding to local curvature. These internal swirls are not particles. They do not exist independently. They have no identity outside the proton. They are simply patterns within patterns—sub-ripples inside the master ripple-knot.

Mainstream physics calls them "quarks," but quarks were a mathematical requirement, not an observable entity. In Unified Energy Dynamics (UED), these sub-ripples are:

- localized curvature zones
- phase-shift pockets
- tension concentrations
- internal oscillation centers

The proton's stability comes from the way these zones balance each other's tension. It is stable because:

1. Its internal energy is high but perfectly locked.
2. Each sub-ripple reinforces the others.
3. The overall curvature reaches equilibrium, not collapse.

This makes the proton extraordinarily stable—so stable that most protons in the universe will outlive stars, planets, and even galaxies. The universe found a way to condense energy into a durable, self-sustaining structure that could survive billions of years. Not bad for a ripple.

5.2 — Quarks: The Most Persistent Misunderstanding in Physics

Quarks were never seen. Never isolated. Never detected directly. Never touched, trapped, or tracked. We inferred them because the proton's internal ripple structure behaved as if something inside it carried fractional properties.

But fractional charge is not a property of a particle. It is a property of internal asymmetry. Imagine pressing your fingers into a stretched fabric. Three points of pressure create curvature zones around them—each zone carrying part of the tension. The tension is real, the curvature is real, the division of tension is real—but the "three objects" are not.

Similarly, the proton's internal structure creates zones where the field behaves as if the charge were divided. Those "quarks" are simply positions within the field where curvature is intensified.

The so-called "strong force" that supposedly holds quarks together is nothing more than:

- the SVE field resisting separation
- ripple tension skyrocketing when symmetry is disturbed
- the architecture reinforcing itself to avoid collapse

This explains confinement perfectly. You cannot pull the internal sub-ripples out because they were never independent entities. Trying to separate quarks is like trying to remove the center of a whirlpool from the whirlpool. The pattern is the object—not its imaginary pieces.

5.3 — Gluons: Ripple Tension Misinterpreted as Particles

In the particle worldview, gluons are imagined as messenger particles carrying force between quarks. But the proton's internal behavior does not require a messenger—only ripple tension.

What the Standard Model describes as "gluon exchange" is simply the field adjusting tension to maintain the integrity of the tri-ripple architecture. When the internal sub-ripples oscillate, tension waves shoot across the structure, balancing curvature and ensuring coherence.

These tension waves appear as energy packets in high-energy scattering experiments, so physicists labeled them "gluons." Yet gluons are never isolated, never leave the proton, cannot exist independently, cannot travel outside the composite ripple, and have no stable identity.

They are not particles. They are ripple-tension updates within the proton's internal architecture.

Imagine tightening a drumhead. The tension shifts across its surface as one region adjusts to another. The shift looks like something traveling, but it is not an object—it is a redistribution of tension. Gluons are exactly that.

The strongest "force" in the universe is simply the strongest tension-equilibrium mechanism inside the densest stable ripple-knot.

5.4 — The Neutron: Stability Through Environment

The neutron is almost identical to the proton, but with a key difference: its internal sub-ripples do not lock perfectly. In free space, the neutron is unstable and decays. But inside an atomic nucleus, it becomes stable—sometimes for billions of years.

Why? Because the neutron is the first structure whose stability depends on external SVE density. In other words, the neutron is stable only when "pressed" by the field gradients created by nearby protons.

Inside a nucleus, the neutron's internal curvature is reinforced by the overlapping ripple fields around it. The nuclear environment creates coherence that does not exist in isolation.

This explains why free neutrons decay in minutes, why neutrons inside nuclei do not, why certain isotopes are stable while others are not, and why nuclear binding is not a force but a coherence phenomenon.

The neutron is like a sculpture of tension that must be held from the outside. Remove the support, and it collapses into simpler ripples—electron + proton + field energy adjustments.

This also disproves any notion that neutrons contain secretly hidden particles. Their stability emerges from environmental coherence, not internal ingredients.

5.5 — Nuclear Coherence: Why Nuclei Hold Together

In the particle worldview, nuclei are said to be held together by a "strong nuclear force." But what we actually observe is:

- no force carriers leaving the nucleus
- no discrete binding particles
- no independent quarks
- no gluons traveling between protons and neutrons
- no direct evidence of attractive glue

The stability of a nucleus comes from ripple coherence. When multiple protons and neutrons come together, their SVE distortions overlap. This overlapping creates shared curvature, balanced tension, smoother gradients, and collective stability.

This is not a force. It is mutual coherence.

Think of dancers locking into a rhythm. Each dancer stabilizes the others through timing, positioning, and feedback. The dance is stable—not because of ropes or glue, but because of synchronization. A nucleus is a dance of ripple-knots whose patterns reduce overall tension when arranged properly.

This framework explains binding energy, fusion, fission, isotopic stability, nuclear shells, decay patterns, and energy release—without introducing fictional carriers or mysterious forces. Nuclear stability is simply the field's preference for lower-tension configurations. Matter works because the universe likes equilibrium.

5.6 — Why the Proton & Neutron Give Matter Its Weight

Mass in UED is not an intrinsic property. It is ripple density.

Protons and neutrons are the densest stable ripples the universe can sustain without collapse. The proton's internal architecture traps enormous amounts of ripple tension. This locked energy creates inertial resistance, which we interpret as mass.

The neutron, being slightly heavier, stores even more internal curvature, but only remains coherent in the right field environment.

In simple language:

- Protons = stable density knots
- Neutrons = pressure-stabilized density knots

Together, they build the atomic nucleus—the field's grand attempt to create long-lasting structures from ripple logic alone.

Without these knots of tension, the universe would have no atoms, no chemistry, no solids, no planets, no stars, no life. All because one day, the field learned how to lock three sub-ripples into a stable architecture.

Now that we have explored the densest stable structures, we turn to their opposite: light.

Where electrons, protons, and neutrons are ripple-knots that lock, light is a ripple that refuses to condense at all. It is movement without mass, propagation without structure, energy without confinement.

In the next chapter, we explore the beauty of pure alignment—how the field passes information without ever creating an object.

Chapter 6 awaits: "Photons: The Freedom of Alignment."

Chapter 6 — Photons

The Freedom of Alignment

Light as pure propagation without condensation

If you toss a stone, it slows and falls. If you throw a ball, it curves and drops. If you fire a rocket, it burns fuel and eventually stops.

But light?

Light never gets tired. Never slows. Never bends unless the field itself bends. Never clings to matter. Never settles. Never condenses. Never dies.

It is the universe's most honest traveler—moving through everything, belonging to nothing, touching all things without being touched in return.

Why? Because a photon is not a particle at all. It is not an object moving through space. It is not a tiny bullet of energy. It is not a wave trapped inside a mystical duality.

A photon is pure field alignment—a ripple that never becomes a knot, never traps itself, never locks its curvature into stability. It is the simplest form of motion in a continuous universe: a disturbance that travels without ever becoming a thing.

In this chapter, we reveal how the photon's freedom explains its constancy, its speed, its dual behavior, and its profound role in everything from warmth to vision to stars.

Light is not an object. Light is the universe in motion.

6.1 — Light Without Edges: The First Clue We Ignored

In the old worldview, light was confusing. Sometimes it behaved like a wave—interference, diffraction, polarization. Sometimes it behaved like a particle—photons ejecting electrons, discrete energy packets, clear signatures in detectors.

Because it refused to pick just one identity, physicists invented the term *wave-particle duality*—a conceptual bandage on a deeper misunderstanding.

But the strange behavior of light was not evidence of complexity. It was evidence of simplicity. Light is neither a wave nor a particle. It is what a continuous field produces when a disturbance does not condense.

To understand light, imagine the SVE field as a calm ocean. A boat creates a swirl (like an electron). But if a pebble taps the surface lightly, a ripple spreads outward freely. It does not condense into a swirl. It does not trap energy. It does not loop back into itself. It simply moves.

That is light: a propagating alignment of the field—nothing more, nothing less.

When light passes, the field temporarily oscillates and then returns to perfect stillness. Light leaves no residue. No knots. No lingering architecture. It is the purest expression of a universe that prefers continuity over fragmentation.

6.2 — The Photon Is Born from Disturbance, Not Emission

In textbooks, we are taught that atoms "emit photons," as if some tiny object shoots out of one place and lands in another.

But what actually happens? An electron—our stable ripple-knot—shifts from one alignment pattern to another. The field responds instantly, propagating the change outward as a traveling ripple of alignment.

The photon is not emitted. The photon is created everywhere along its path as the field updates itself.

Picture this: you pluck a guitar string. The vibration moves along the string, but the motion at one point does not physically travel as an object. The disturbance is recreated at every point in sequence.

That is light. Not a bullet crossing space, but a reconfiguration of the field propagating from point to point.

When the alignment reaches a surface—your eye, a sensor, a wall—the field adjusts, and energy appears to "arrive." We mistake this arrival for a particle hitting a target. But the universe was never sending projectiles. It was sending updates.

The photon is a moving change, not a moving thing.

6.3 — Why Light Always Travels at One Speed

Nothing in the universe behaves with such consistency as light. Whether from a candle or a quasar, whether traveling through vacuum or gently passing through glass, light always respects one truth: it moves at the maximum speed allowed by the field.

But why is this speed universal? Because a photon has no mass, no locked curvature, no internal sub-ripples, no density knot, no inertia. There is nothing to slow it down. Nothing to resist propagation. Nothing to negotiate with.

A photon is simply the field propagating its own alignment at the fastest rate the medium allows. Like the tension on a perfectly tight string, the propagation is automatic and universal.

Changing the speed of light would require changing the structure of the SVE field itself. This is why time dilates near gravity, length contracts, and moving observers see different realities—yet the speed of light remains untouched.

Light is not a particle obeying a speed limit. It is the field revealing its fundamental rhythm. The universe dances at one beat—and that beat is *c*.

6.4 — Light as Pure Alignment: The End of Wave–Particle Duality

Wave–particle duality was never a truth—it was a compromise. Light only appears wave-like because the field oscillates. It only appears particle-like because interactions happen in discrete events. But neither observation tells us that a tiny thing must exist.

Let us simplify:

- **Why light behaves like a wave** Because the field oscillates as the disturbance propagates.
- **Why light behaves like a particle** Because energy exchange happens in crisp transitions when the field locks or unlocks around electrons.

When the alignment reaches a detector, a single event occurs. Our sensors interpret that event as "a photon hit." But what actually happened was:

- the field realigned
- energy was delivered in a discrete jump
- electrons shifted their configuration
- a ripple ended its journey

The event is discrete. The movement is continuous. No paradox. No duality. Only the behavior of a continuous field interacting with stable ripple-knots.

Light behaves simply because the field behaves simply. It was our fragmented thinking that made light look contradictory.

6.5 — Visualizing Photons Without Particles

Photons confuse us because we try to imagine them as objects. But try this instead:

Imagine a ripple traveling across a pond. The water at each point moves up and down. The ripple travels outward. But the ripple is not a thing. It is a movement.

Or imagine a stadium wave. The wave travels around the stadium, but the people do not travel. The motion is recreated at each point. This is exactly how the photon moves—a disturbance propagated, not an object transported.

Or imagine a shift of color across a LED strip. The "movement" of color is not one LED moving. It is a sequential update.

Light is the universe updating itself—not throwing particles.

Once this picture settles, every paradox dissolves. Light never needed a particle identity. It never had one. The photon is the universe at its most free—a ripple refusing to become matter.

6.6 — Why Light Interacts So Cleanly with Matter

Electrons are stable ripple-knots. Protons and neutrons are dense ripple architectures. Photons, being pure alignment, interact with these structures in clear, predictable ways:

- **Light can push electrons** — not by impact, but by changing the local field alignment that the electron knot depends on.
- **Light can eject electrons** — not because a particle hits them, but because the ripple pushes the electron out of its stable configuration.
- **Light can heat matter** — not by delivering tiny bullets of energy, but by disturbing the local curvature and increasing ripple agitation.
- **Light can be absorbed** — not because it is swallowed, but

because its alignment is used to restructure internal patterns of matter.

- **Light can be reflected** — not because it bounces, but because the surface rearranges the propagating alignment.

Everything makes more sense when light is not a thing but a behavior.

We have now seen the two extremes of ripple behavior:

- **Electrons** — ripples that lock and become matter
- **Photons** — ripples that propagate freely

Now we move to the middle ground—to the most misunderstood structures in modern physics: quarks and gluons. Not particles. Not building blocks. Not tiny imprisoned objects. But internal sub-ripples misinterpreted as entities because we insisted on seeing everything as particles.

In the next chapter, we walk into the heart of the proton to reveal the true nature of those mysterious "quarks."

Chapter 7 awaits: "Quarks & Gluons: The Greatest Misunderstanding."

Chapter 7 — Quarks & Gluons

The Greatest Misunderstanding

Internal distortions mistaken as particles

Stand in a hall of mirrors, and you will see a dozen versions of yourself. None of them are real. None of them exist independently. They are simply distortions—patterns created by angles, surfaces, and reflections.

But imagine if someone walked into that hall, saw the reflections, and declared: *"Look! A new species!"*

That is exactly what we did with quarks and gluons. We looked inside the proton—saw internal distortions, tension pockets, and ripple behaviors—and mistook them for tiny particles with exotic properties. We gave them names, equations, charges, spins, and built a massive mathematical structure around them.

But we never found one. Not a single quark. Not a single gluon. Not once, not anywhere, not in any experiment. Because they were never there.

This chapter dismantles that illusion and reveals the truth hidden in plain sight: quarks and gluons are not entities. They are internal ripple distortions of a deeper field architecture.

Once we see this clearly, the entire "strong force paradigm" collapses, and a far more elegant picture emerges—one without confinement mysteries, without color charges, without messenger particles, without paradoxes.

Just one field, performing internal dances that we misunderstood for decades.

7.1 — The Mistake Begins: A Proton Is Not Empty Inside

Textbooks often show protons as solid spheres containing three little balls inside: one blue, one red, one green—the famous "quark triplet." It looks neat, familiar, reassuring. It feeds our intuition that matter is built of smaller pieces, like Russian dolls or Lego bricks.

But this picture is fiction.

When we probed the proton with high-energy scattering, we did not find particles inside. We found uneven energy densities, tension concentrations, swirling regions of curvature, asymmetrical ripple pockets, and dynamic, continuously shifting internal patterns.

These signatures told us something was happening inside—but they did not tell us what it was. Instead of considering a field-based explanation, physics jumped to the most classical idea possible: *"If something inside the proton creates different responses, it must be smaller particles inside it."*

Thus quarks were born—not from observation, but from interpretation.

We mistook patterns for constituents. We saw effects and invented causes. We saw distortions and imagined creatures.

7.2 — What We Call "Quarks" Are Actually Sub-Ripple Curvature Zones

In a continuous SVE field, a composite ripple-knot like the proton naturally forms internal curvature zones—regions where the field twists, bends, or compresses to maintain overall stability.

These zones act like anchors or nodes inside the structure:

- They carry partial tension
- They shape the overall swirl
- They distribute curvature
- They prevent collapse
- They maintain symmetry

In UED and RFD language: **Quarks = internal field distortions within the proton's ripple architecture.**

They do not exist as objects. They do not travel. They do not have independent identity. They cannot be separated. They cannot be extracted. They cannot exist outside the proton.

Why? Because they were never things in the first place.

Try removing the center of a whirlpool. Try extracting the bend from a wave. Try isolating one end of a rainbow. Impossible.

Quarks are structural positions within a larger ripple—not fundamental particles.

This is why every attempt to isolate a quark failed catastrophically. Physics expected prisoners. But there were no prisoners. Only patterns.

7.3 — Confinement: The Paradox That Should've Ended the Particle Model

Quark confinement is often presented as one of the great triumphs of quantum chromodynamics (QCD). But in reality, it should have been the final nail in its coffin.

Confinement states:

- quarks cannot be isolated
- the force between them increases as they separate
- the energy grows without limit
- pulling them apart creates more quark pairs
- the universe will never let you see a quark

This is absurd for a particle theory. Imagine a theory of birds that says: "*You can never hold just one bird. If you try, the sky creates more birds.*" Absurdity is often a clue that something fundamental is wrong.

The field explanation is simple: you cannot pull a sub-ripple out of the ripple because it is not a distinct entity. Separation does not create energy—it distorts the field so violently that the field reorganizes into new stable knots (electrons, protons, mesons, etc.). We interpreted that reorganization as "quark pair creation." But the universe was simply repairing the damage to the field.

Confinement is not a force. It is the natural resistance of a stable ripple architecture against being torn apart. The proton is not a container. It is a behavior. You cannot take the pieces out—because there are no pieces.

7.4 — Gluons: Tension Waves Misread as Messengers

Gluons were introduced to explain why quarks seemed to exchange energy so rapidly. Again, it looked like something was traveling between the "quarks," so physics assumed: "*There must be a messenger particle.*"

Thus gluons were created—eight of them, in fact—each carrying "color charge" in a mathematically elegant but physically invisible system.

But what are gluons really? Nothing but ripple-tension waves within the proton's internal architecture. When the proton's curvature adjusts, tension redistributes rapidly across the structure. This redistribution creates signatures that look like energy packets in detectors. So we labeled them "gluons."

But gluons:

- have never been isolated
- do not appear as free objects
- exist only as mathematical entities

- emerge only inside the proton
- dissolve instantly when the structure collapses

Because they are not particles. They are simply the field updating its internal tension.

Imagine tightening a drumhead. When you adjust one side, ripples of tension spread across the surface. You could imagine tiny messengers carrying the force, but nothing is traveling—it is the surface itself adjusting.

The proton is the drumhead. The gluons are the adjustments. The universe was telling us the truth, but we kept insisting on particles.

7.5 — The Illusion of Parton Models and Deep Scattering

Deep inelastic scattering experiments in the 1960s and 70s are often cited as "evidence" for quarks. But a closer look reveals that these experiments measured uneven momentum distribution, spikes in field density, local curvature concentration, unstable resonance peaks, and internal tension responses. In other words: the internal structure of a ripple—not constituents of matter.

Feynman himself called these patterns *partons* and explicitly warned they should not be assumed to be particles. Yet the scientific community preferred the particle interpretation because it was easier to imagine. Thus quarks became "real" not because we observed them, but because we needed something to map the mathematical spikes to.

Every deep scattering phenomenon is fully explained when the proton is understood as a composite ripple architecture:

- high-energy probes disturb sub-ripples
- local curvature spikes respond
- tension waves propagate
- energy redistributes

- detectors record discrete events

Nothing here proves the existence of objects. Everything here proves the existence of patterns.

7.6 — Why Quarks Can't Exist Outside Nucleons

In every theory that uses particles, "free quarks" should be possible at sufficiently high energies. But after trillions of collisions and unimaginable energy densities, no experiment has ever isolated a single quark.

This is not a mystery. It is a clue.

The moment you try to extract a quark:

- the field loses coherent curvature
- the ripple architecture collapses
- tension spikes reorganize
- new particle signatures appear
- the proton ceases to exist

In UED terms: a quark cannot exist outside the proton because the field cannot sustain a sub-ripple without the full architecture.

Pulling a quark out is like trying to remove the "north pole" of a magnet. You cannot isolate the pole because the pole is a property of the whole.

Quarks are not building blocks. They are geometric consequences of the proton's locked ripple structure. Nothing more. Nothing less.

7.7 — What This Correction Means for Physics

Recognizing quarks and gluons as internal distortions, not particles, dissolves decades of confusion. It explains:

- why quarks cannot be isolated

- why gluons never appear freely
- why proton internal energy is so high
- why the proton is extraordinarily stable
- why neutrons are stable only inside nuclei
- why deep scattering produces complex signatures
- why QCD calculations are notoriously difficult
- why confinement is unavoidable
- why nuclear forces behave strangely
- why high-energy collisions create "jets"
- why hadrons always emerge from collisions

Every paradox dissolves. Every mystery becomes geometric. The universe becomes simpler—not because we removed complexity, but because we removed imaginary entities.

Once we stop insisting on particles, a coherent picture emerges. Protons and neutrons are not collections of pieces. They are dynamic ripple architectures performing internal dances to maintain stability. Quarks are not prisoners. Gluons are not messengers. The strong force is not a force. Everything is the field holding itself together.

Now that the illusion of quarks and gluons is gone, we face the next question: what about the short-lived "particles" produced in high-energy collisions—the W and Z bosons, the Higgs, the tau, the exotic resonances? Are these real objects, or are they yet another layer of misunderstanding?

In the next chapter, we look at the heavy modes—the brief flickers of intense curvature that appear only for a moment before dissolving back into the field.

Chapter 8 awaits: "The Heavy Modes: W, Z, Higgs & the High-Energy Flickers."

Chapter 8 — The Heavy Modes

W, Z, Higgs & the High-Energy Flickers

Short-lived distortions that cannot lock

Strike a tightly stretched drum hard enough, and for a fraction of a second the surface contorts—wild peaks, sharp dents, chaotic folds—shapes that exist only while the force is extreme. The moment the impact fades, the drumhead relaxes back into its stable geometry.

Now imagine the SVE field as that drum. High-energy collisions are the strike. And those bizarre shapes—the contorted, short-lived forms—are what physicists call W bosons, Z bosons, Higgs bosons, taus, deltas, lambdas, exotic resonances.

Not particles. Not building blocks. Not real constituents of nature. Just violent distortions of the field that exist only under extreme tension.

This is the story of heavy modes—brief, unstable flickers of curvature born in environments so intense that the field temporarily loses its calm and forms ripples it cannot sustain.

They live for 10^{-25} to 10^{-27} seconds. They never travel. They never stabilize. They never show structure. They appear and vanish like sparks when a metal grinder cuts steel.

This chapter explains why heavy modes exist, why they decay instantly, why they cannot lock like protons or electrons, and why viewing them as "particles" has misled physics for half a century.

8.1 — What Are Heavy Modes? The Universe Under Extreme Tension

To understand heavy modes, we must begin with the idea of ripple locking. Some configurations of the SVE field—like electrons, protons, and atoms—manage to stabilize into long-lived geometries. They are symmetrical, balanced, and their internal curvature can be held indefinitely.

But heavy modes fail this test. Their curvature is too intense, their tension too high, their symmetry too fragile. They flicker into existence only when immense energy is forced into a tiny volume of space.

Think of it this way:

- A proton is a stable knot.
- A W boson is a rip in the knot.
- A Higgs is a massive bulge in the field.
- A tau lepton is a curvature overshoot.
- Exotic mesons are unbalanced swirl attempts.

None of these can survive. The field does not support them. As soon as they form, they collapse—unwinding themselves back into lighter, stable configurations. This collapse is what physicists call "decay."

The heavy mode was never a true structure. It was a field tantrum.

8.2 — Why Heavy Modes Require Enormous Energy to Create

Why does the Large Hadron Collider need energies millions of times larger than chemical processes to create these heavy flickers? Because heavy modes are not natural states. They are forcefully imposed distortions.

To create a W boson, Z boson, or Higgs, the field must be compressed so violently that:

- curvature overshoots
- symmetry breaks
- density spikes
- tension grows beyond stability
- the ripple momentarily folds

This requires:

- titanic pressure
- extreme collision energies
- confinement in microscopic volumes
- field gradients far beyond normal

It is like trying to hold a whirlpool together in the middle of a tsunami. You can create strange shapes, but they will not last.

Mainstream physics sees these flickers as fundamental carriers, critical particles, mediators of forces, ingredients of the universe. But the universe itself never uses these modes naturally. They appear only in the lab when we abuse the field.

Nature does not create massive bosons. Nature creates stability. The heavy modes are byproducts of violence, not fundamental features of existence.

8.3 — The W and Z Bosons: Failed Attempts at Locked Curvature

The W and Z bosons are often celebrated as triumphs of particle physics. Yet their properties tell a completely different story. They have enormous effective mass, extremely short lifetimes, no internal structure, no capacity to exist freely, no stability in vacuum, no ability to propagate, and no role outside extreme conditions.

In a field-based view: *W/Z bosons are unstable curvature spikes created when the proton's architecture collapses locally under stress.* They are not messengers of the weak force. They are not carriers of anything. They are simply what the field looks like when it is pushed beyond its structural limits.

Visualize a trampoline:

- A child jumps gently → stable ripples
- Children collide mid-air → chaotic distortions
- For a split second → the surface forms extreme peaks

Those peaks are W/Z equivalents. They appear only under impact. They collapse instantly. They are not entities. They are stress responses.

Decay is not a process. Decay is the field returning to sanity.

8.4 — The Higgs Boson: The Most Misunderstood Flicker

The Higgs is often portrayed as the "particle that gives mass to the universe." But its behavior tells another story. The Higgs appears only in extreme collisions, vanishes instantly, does not exist freely, cannot propagate across space, does not show up in any natural environment, has no stable geometry, and has no structural meaning outside the lab.

In UED/RFD: the Higgs is a massive, symmetric bulge of ripple density formed when the field is overstressed. It is the ultimate "over-compression"—a moment where the field collapses inward, trying to absorb too much curvature in too small a space.

It does not give mass. Mass is locked curvature, not Higgs interaction. The Higgs is simply what the field becomes when overfed with energy.

Calling it a particle was the greatest misinterpretation of the century. The Higgs is not a source. It is a symptom.

8.5 — Why Heavy Modes Decay Instantly

Every heavy mode has a lifetime between 10^{-25} and 10^{-27} seconds. This is not a coincidence—it is a limit.

When the field exceeds its structural capacity, it cannot maintain the curvature spike. The collapse is immediate:

- internal tension unwinds
- curvature relaxes
- energy radiates outward
- stable ripple configurations emerge

The field always chooses the path of least tension.

Decay patterns are not evidence of hidden components. They are evidence of the field restoring coherence. Heavy modes die because the universe is allergic to them. They do not lock. They cannot lock. They were never meant to lock.

8.6 — Why High-Energy Colliders Produce "Jets," Not Particles

In particle physics, the jets seen in detectors are assumed to be showers of outgoing particles. But jets are actually:

- ripple fragmentation patterns
- field relaxation pathways
- symmetric energy shedding
- alignment waves
- proton remnants reorganizing

When a heavy mode collapses, its energy disperses geometrically—as cones, sprays, or lines. Detectors record these patterns as discrete hits and reconstruct them backward into "particles."

But what appears as a list of exotic particles is simply one violent event relaxing into many stable or semi-stable ripples.

The jet is the decay, not the particles. It is the signature of the field returning to order.

8.7 — Heavy Modes Have No Role in Natural Physics

Tour the universe: atoms, molecules, planets, stars, life—none require W bosons, Z bosons, or Higgs bosons to function. Not one natural process depends on their presence.

Why? Because heavy modes are not real structures. They are distortions we force the field into using extreme artificial conditions.

The universe prefers:

- electrons
- protons
- neutrons
- photons

The universe does not use W, Z, Higgs, tau, delta, sigma, omega... These flickers exist only in human-built machines. They are field screams, not fundamental whispers.

8.8 — What This Means for the Future of Physics

Recognizing heavy modes as distortions, not particles, unlocks a simpler, clearer, more intuitive physics. It explains:

- why heavy modes don't appear naturally
- why they decay instantly
- why they require enormous energy to create
- why QCD and electroweak models feel "patched"
- why unification attempts struggle
- why mass doesn't need a Higgs field

- why the Standard Model looks like a forced taxonomy

More importantly, it shows that: *The universe is not built from hundreds of species of particles. It is built from a handful of stable ripple behaviors and a spectrum of unstable ones.*

This unifies physics without inventing imaginary entities. It simplifies equations without losing accuracy. It dissolves paradoxes without creating new ones.

Now that we understand the heavy, unstable flickers of the field under stress, we must look at the opposite extreme: neutrinos—the lightest, faintest whisper the field can produce.

If heavy modes are over-compression, neutrinos are under-formation. Almost nothing, yet still something. They pass through planets, stars, and galaxies as if matter doesn't exist—because most of matter, at its core, is the field they glide through effortlessly.

In the next chapter, we explore how minimal distortions travel enormous distances and leave faint but undeniable signatures.

Chapter 9 awaits: "Neutrinos: The Almost-Nothing That Still Leaves a Trace."

Chapter 9 — Neutrinos

The Almost-Nothing That Still Leaves a Trace

Minimal distortions, maximal travel

Imagine a whisper so soft that it passes through walls, mountains, oceans, and planets without disturbing a single grain of matter. Imagine a ripple so gentle that fifty trillion of them pass through your body every second—and you never feel a thing. Imagine a disturbance so faint that it requires detectors the size of buildings, buried deep underground, just to catch a few whispers each day.

This is the neutrino—the universe's quietest traveler, the faintest signature of motion, the ripple that barely exists yet still leaves a trace.

But what is a neutrino really? Not a particle. Not a point-like object. Not a mysterious ghost-ball slipping through matter.

A neutrino is the smallest possible deviation the SVE field can sustain as a traveling ripple—a ripple so minimal, so low in curvature, so close to stillness that almost nothing interacts with it.

It is the universe's proof that even near-zero distortions still obey ripple logic.

This chapter explores how such a faint disturbance can exist, why it travels cosmic distances without fading, why it oscillates between "types," and what its existence reveals about the structure of the field itself.

9.1 — The Lightest Ripple: What a Neutrino Actually Is

Electrons are stable knots. Protons and neutrons are dense architectures. Photons are pure alignment waves. Heavy modes are violent flickers.

Neutrinos sit on the opposite end of the spectrum—the least intrusive, least condensed, least reactive ripple the field can support.

In UED/RFD terms: a neutrino is a barely-formed curvature mode that never reaches full locking. It is not massless like a photon, but its mass is so tiny that it behaves like a near-free wave.

It has:

- minimal curvature
- minimal tension
- minimal density
- minimal asymmetry
- minimal interaction footprint

It is the softest ripple the field can create without dissolving instantly into pure alignment.

This makes the neutrino neither a particle nor a photon, but a strange in-between state: a traveling curvature that barely exists, yet travels with astonishing persistence. The field allows it because the distortion is too weak to trigger collapse or stabilization.

9.2 — Why Neutrinos Barely Interact with Matter

Mainstream physics explains neutrino invisibility as "weak-force interaction only." But the reality is much simpler: neutrinos rarely interact because their curvature is too small to disturb stable ripple-knots like protons and electrons.

Matter is made of dense field knots—structures with locked curvature. A neutrino passing through such knots is like a gentle breeze passing through a forest:

- It flows around obstacles.
- It bends slightly but keeps going.
- It does not break, push, or disrupt.

Interaction would require the neutrino's minimal curvature to cause local field rearrangement, but in most cases:

- its amplitude is too tiny
- its tension too weak
- its asymmetry too mild

Thus it passes through atoms, molecules, human bodies, planets, stars, and galaxies as if they are not even there. Only once in billions of encounters does a neutrino disturb the field enough to trigger a noticeable event.

This is not because neutrinos are elusive particles. It is because they barely exist as curvature.

9.3 — Why Neutrinos Oscillate: The Field's Whisper Changing Shape

Neutrino oscillation—their ability to change "flavor" (electron → muon → tau neutrino) as they travel—is one of the strangest behaviors in physics. Yet the field interpretation is elegant:

A minimal ripple cannot maintain a fixed internal pattern because it is too weak to lock into one shape.

Electrons and protons hold their geometry forever. Heavy modes collapse instantly. But neutrinos sit on the threshold:

- not stable enough to maintain identity

- not unstable enough to dissolve
- just coherent enough to travel
- just weak enough to fluctuate

This creates natural phase drift in the ripple's internal structure. As the neutrino moves, tiny differences in curvature reconfigure, and the ripple takes on one of several permissible patterns.

The "flavors" are not different species—they are different modes of the same weak ripple.

It is like watching a faint light pass through fog:

- its shape stretches
- its color shifts
- its edges flicker
- its brightness fluctuates

But it is still the same light.

Neutrino oscillation is not magic. It is the geometry of near-instability.

9.4 — Why Neutrinos Travel Cosmic Distances Without Fading

One of the mysteries of neutrinos is their almost perfect endurance. A ripple so weak should dissolve quickly. Yet neutrinos survive journeys from exploding stars, distant galaxies, and even the early universe.

How? Because weak ripples are less vulnerable to field disturbances.

High-curvature ripples—electrons, protons, heavy modes—are sensitive:

- to collisions
- to gradients
- to curvature stresses

- to local field distortions

But minimal curvature ripples glide effortlessly along the SVE landscape. The field treats them gently—they don't provoke reactions and therefore don't suffer from them.

A neutrino is the universe's quiet traveler because it stays below the threshold that causes friction-like interactions. It rides the field like a leaf drifting on a calm stream—too light to sink, too faint to affect the flow.

9.5 — Visualizing Neutrinos: The Ripple That Barely Wiggles

To grasp the neutrino, don't imagine a ball or a particle. Think instead:

A tiny wrinkle on a massive sheet—you can run your hand across it and barely feel a thing. A slight shimmer on a lake's surface—almost invisible even when you look directly. A whisper of wind through a valley—it moves endlessly but disturbs little. A ghost image in a hologram—there, yet not fully formed.

These analogies show why neutrinos behave like ghosts:

- They are too gentle to cause disruptions.
- Their curvature never locks.
- Their tension is minimal.
- Their travel requires almost no energy.
- They fade only when they encounter a rare perfect alignment that triggers interaction.

Neutrinos feel supernatural only because we insist on imagining particles. Once we see ripples, they become beautifully simple.

9.6 — Neutrinos and the Early Universe: The First Free Travelers

In the early universe:

- electrons locked early
- protons formed soon after
- photons were scattered constantly
- matter thickened into clouds

But neutrinos? They passed through everything, untouched. They became the first free travelers—ripples set loose in a dense, chaotic cosmos.

Their minimal curvature allowed them to escape long before photons could travel freely.

Thus:

- neutrinos carry information from earlier stages of the universe than any other messenger
- they are remnants of cosmic infancy
- their presence reveals the earliest field conditions
- they teach us how the SVE stabilized as the universe cooled

In this sense, neutrinos are not ghosts: they are witnesses. They are signatures of a universe just beginning to learn stability.

9.7 — Why Neutrinos Matter to the Field Picture

Understanding neutrinos strengthens the SVE/RFD worldview in three profound ways:

1. **They prove that the universe supports a full spectrum of ripple intensities.** From pure alignment (photons) to locked knots (electrons), to dense architectures (protons), to heavy

flickers, to faint ripples (neutrinos).

2. **They show that identity is not fixed but shaped by curvature stability.** Neutrino oscillation is the clearest demonstration that "particles" are behaviors, not objects.
3. **They reveal that matter is not needed for propagation.** Neutrinos remind us that the field itself is the medium of existence. Matter is optional. The field is fundamental.

With neutrinos, we have now covered the full spectrum of ripple behaviors:

- locked
- semi-locked
- propagating
- collapsing
- minimal

Every so-called "particle" falls into one of these field categories.

Part II reconstructed matter from the field up. Now, in Part III, we turn our focus outward—to how these ripple structures interact, attract, bind, repel, and shape the world we experience.

Interlude — When Matter Loses Its Edges

A reflection at the end of Part II

There is a quiet moment in every great scientific shift—when the old idea dissolves, but the new one has not yet fully taken shape.

That is where we stand now: on the far shore of Part II, looking back at everything we once called "matter," and realizing how much of it was only a habit of imagination.

For over a century, we believed matter was built from pieces—electrons, protons, neutrons, quarks, gluons, leptons, bosons... a zoo of fragments stitched into handbooks and drilled into generations of students.

But in Part II, we watched those fragments soften, blur, merge, and finally dissolve into continuity.

- Electrons stopped being particles and became stable ripple-knots.
- Protons stopped being bagged quarks and became tri-curvature architectures.
- Neutrons stopped being neutral puzzles and became environmentally stabilized patterns.
- Photons stopped being packets and became pure traveling alignment.
- Heavy bosons stopped being messengers and became

momentary stress distortions.

- Neutrinos stopped being ghost particles and became faint whispers of curvature.

One by one, every resident of the particle zoo returned to the field where it truly belonged.

Not as things. As behaviors. Not as ingredients. As patterns. Not as building blocks. As ripple configurations created, shaped, and dissolved inside a single continuous SVE substrate.

The more we looked, the fewer particles we found. And the fewer particles we found, the more coherent the universe became.

No dualities. No paradoxes. No force carriers. No exotic species. No hidden machinery.

Just the field—doing what the field does.

Matter is no longer a collection of objects. It is a vocabulary of stable rhythms. A library of locked geometries. A spectrum of persistence within one medium.

This changes everything.

When matter loses its edges, forces lose their carriers. Interactions lose their intermediaries. Mass loses its mystery. Energy loses its disguise. And physics loses the illusion that the universe is built like a machine.

What remains is something simpler, cleaner, far more elegant:

A universe where nothing is separate. A universe where everything is ripple. A universe where matter and motion are two dialects of the same language.

Part II did not just rewrite matter. It rewrote our expectations. It showed us that the universe is unified by nature, fragmented only by interpretation.

Ahead lies the next challenge—perhaps the most radical one:

If there are no particles, then there are no force carriers. If there are no carriers, then interactions are not exchanges...

They are consequences. Emergent. Inevitable. Geometric.

In Part III, we follow those consequences—to see how the universe pulls, pushes, binds, repels, and shapes the grand dance without ever invoking a single object.

We leave behind the zoo, and enter the field's pure logic.

The vista is wide open now. The walls are gone. The floor is continuous. And the field is waiting.

PART III — THE DEATH OF FORCE

"Nothing pulls. Nothing pushes. The field only seeks balance— and we mistake its corrections for force."

Short Overview of Part III

Part III marks the most radical step in the book: we dismantle the entire concept of force—a pillar of classical and modern physics—and replace it with something far simpler, far more universal, and completely aligned with the behavior of a continuous field.

If Part I ended the illusion of particles, and Part II reconstructed matter as ripple architectures, then Part III reveals the final truth:

The universe does not use forces. It only uses consequence.

Every attraction, every repulsion, every orbit, every fall, every bond, every field line, every curvature—all emerge from the same mechanism: *ripple tension interacting with SVE gradients.*

This part begins by showing why the classical idea of force—a push or pull transmitted by objects or carriers—never reflected reality. What we observed were simply the field's attempts to relax curvature, align distortions, and restore equilibrium.

Then we reinterpret the most sacred phenomena in physics—electricity, magnetism, gravitational attraction—through the lens of field consequence.

- Atoms bind not because forces act, but because ripple architectures find lower-tension configurations.
- Masses attract not because gravity pulls, but because field gradients guide curvature toward stability.
- Charges repel or attract not because they "exert" influence, but because their asymmetries seek tension relief.

Part III replaces a complex, fragmented list of forces with a single unified behavior: *the field minimizing tension.*

What emerges is not only clearer and more intuitive, but also powerful enough to unify phenomena across quantum, classical, and cosmic scales.

By the time Part III ends, the reader will see the universe not as a battlefield of forces, but as a continuous choreography of balance—a world where nothing pushes or pulls, yet everything moves with purpose.

Chapter 10 — Why Force Never Existed

Ripple tension × SVE gradients = consequence

Think of the last time you pushed open a door. Your hand pressed forward, the door swung, your body felt the resistance. Everything about the experience seems obvious: you applied a force, the door moved because of your push—a straightforward chain of cause and effect.

But look closer.

Your hand never touched the door's atoms directly. Electric fields—nothing more—created a barrier. Curvature in the surrounding field resisted the motion. Your body, the door, the air between—all were interacting through invisible ripples. Nothing pushed. Nothing pulled. The field was simply adjusting itself to minimize tension.

You never touched the door. You never pushed it. The "force" you felt was an illusion of interactions inside one continuous field.

This chapter exposes the foundation of that illusion—showing that force was never fundamental, never real, never required to explain motion or interaction.

Everything we called a force is simply consequence: the natural outcome of ripple tension interacting with SVE gradients. Pushes and pulls were human metaphors. The universe never used them.

10.1 — The Classical Error: Assuming Contact Means Cause

Classical physics was built on a simple intuition: objects push and pull each other. Stones fall because Earth "attracts" them. Magnets pull because a "force" acts across distance. Charges repel because "fields exert influence." Springs restore motion because of "elastic force."

These descriptions worked for centuries because they matched our experience. But they were never fundamental. They were conveniences—shortcuts that hid the deeper, continuous behavior of the field.

Here is the truth classical physics missed:

1. No object ever touches another.
2. No push is ever transmitted.
3. No pull ever jumps across space.
4. No force is exchanged.
5. Every interaction is field adjustment—not force mediation.

Consider two charged objects. They do not exert forces on each other. They create tension in the field between them, and the field relaxes by bringing them into a configuration of lower tension. We observe this adjustment as "motion." We call it a force.

But the universe is not listening to those labels. It is just doing what fields do: seeking equilibrium.

10.2 — Ripple Tension: The Real Reason Things Interact

When a ripple exists in the SVE field—whether stable (electron), semi-stable (neutron), or propagating (photon)—it creates localized curvature. Curvature means tension. Tension means imbalance. Imbalance means the field wants to relax.

This desire to minimize tension is the true engine of all interactions.

1. Two electrons repel not because they "push" each other, but because their curvature patterns overlap in a way that increases local tension. The field reduces tension by separating them.
2. A proton and electron attract not because of a magical "electromagnetic force," but because their opposite curvature asymmetries fit together in a lower-tension configuration.
3. A mass curves space not by emitting gravity, but by creating a gradient in field density; other ripples naturally follow the slope of least tension.
4. Light bends around stars not because gravity pulls it, but because the field it travels through is denser near massive bodies.

The universe is not pushing or pulling anything. It is relaxing.

Everything moves because tension wants to go down and alignment wants to go up. What we call motion is consequence.

10.3 — SVE Gradients: The Landscape Everything Must Follow

If ripple tension is the "why," SVE gradients are the "how." The SVE field is not uniform everywhere. It has:

- density gradients
- curvature gradients
- energy gradients
- symmetry gradients
- tension gradients

These gradients act like slopes in a landscape. Ripples naturally follow the path that minimizes tension along these slopes. This is what we misinterpreted as force.

A stone falls not because gravity pulls it, but because the field beneath it has a gradient that guides curvature downward. Magnets attract or repel not because of a magnetic "force," but because their internal swirl patterns reconfigure the surrounding field into low-tension or high-tension zones. Atoms bind not because electrons "jump," but because the combined architecture creates a lower-tension state. Planets orbit the Sun not because the Sun pulls them, but because the SVE gradient curves the path of least resistance into an ellipse.

The universe is an immense field with hills and valleys of tension. Everything moves as consequence of the landscape. We invented "force" to avoid explaining the terrain.

10.4 — Why "Force Carriers" Never Made Sense

To preserve the idea of force in a quantum world, modern physics proposed messenger particles:

- photons for EM
- gluons for strong force
- W/Z for weak force
- hypothetical gravitons for gravity

This patch was elegant mathematically but conceptually incoherent. Problems appeared immediately:

1. Force carriers are never directly observed.
2. They cannot explain instantaneous coherence.
3. They fail in extreme curvature conditions.
4. They require paradoxical identities.
5. They imply discrete exchanges in a continuous universe.

6. They contradict the very notion of fields being fundamental.

The universe was screaming continuity while physics insisted on packets.

But in UED/RFD: there is no need for carriers because there is nothing to carry. Interactions are not transmissions. They are structural adjustments. No force jumps between objects. No messenger delivers influence. No exchange occurs. Everything happens locally because the field is continuous everywhere.

Force carriers were the scaffolding that helped quantum mechanics survive its early confusion. Now that the field is understood, the scaffolding comes down.

10.5 — Why Motion Happens Without Force

This is the most counterintuitive part—because we've grown up believing that nothing moves without a push. But in a field universe:

Motion is the default. Force is the illusion.

A ripple moves when:

- tension gradients guide it
- density slopes redirect it
- field curvature carries it
- coherence encourages alignment

It stops moving when:

- curvature locks
- asymmetry stabilizes
- tension reaches minimum

No push. No pull. No cause. No agent. Only consequence.

Imagine a marble rolling down a hill. You can call gravity a force—or you can say: *"The marble moves because slopes exist."*

Likewise:

- electrons arrange into orbits
- planets settle into their paths
- chemicals bond
- light bends
- galaxies rotate

Not because something forces them—but because the field provides the slopes.

Motion is geometry. Not force.

10.6 — Visualizing Interactions Without Force

To make this intuitive, picture a stretched elastic sheet. Place two heavy spheres on it. The sheet curves. A lighter ball placed nearby rolls toward the heavier one. No force pulled it. It simply followed the curvature.

Now translate that to the SVE:

- dense ripple architectures curve the field
- lighter ripples follow the gradients
- stable patterns emerge naturally
- repulsion and attraction are differences in tension

It's not a push or pull. It's a landscape.

Here's another analogy: imagine a river. A leaf drifts downstream. A stone alters the water's flow. Two currents merge. Nothing exerts force. Everything reconfigures.

The SVE field behaves the same way—exquisite, continuous reconfiguration without any particle intermediaries.

10.7 — Ending the Reign of Force Once and for All

We practiced force for thousands of years as a conceptual crutch. It was a powerful metaphor. It served well. It taught us mechanics, engineering, technology.

But now, physics has grown up. The deeper we looked—into atoms, nuclei, quantum fields, curved spacetime—the more obvious it became:

Force was never fundamental.

The universe does not push. The universe does not pull. It adjusts. It balances. It minimizes tension. It follows gradients.

Force was a story we told ourselves when we did not yet know how to read the field. Now we do.

The final equation of this chapter is simple:

Interaction = Ripple Tension × SVE Gradient Force = human interpretation of that consequence

That is the death of force. And the birth of a unified physics.

If force never existed, and all interactions are geometric consequences, then what becomes of attraction, repulsion, and gravity—the pillars of all physical behavior?

How do we reinterpret the fall of an apple, the orbit of a moon, the binding of molecules, the expansion of space itself?

In the next chapter, we rewrite the world's most familiar interactions through the lens of field behavior—removing force entirely and replacing it with pure consequence.

Chapter 11 awaits: "Attraction, Repulsion & Gravity Reinterpreted."

Chapter 11 — Attraction, Repulsion & Gravity Reinterpreted

Unified field behavior across all scales

Pull your hand toward a magnet. Drop a stone from your balcony. Watch a soap bubble shrink into a perfect sphere. Observe two planets locked in orbit. See a spark jump between two wires.

At first glance, these events look nothing alike. One pulls. One falls. One compresses. One circles. One jumps.

But beneath all these behaviors lies a single, universal truth:

Nothing pulls anything. Nothing pushes anything. The field simply rearranges itself to reduce tension.

Attraction is not a force. Repulsion is not a force. Gravity is not a force. Bonding is not a force. Even the so-called "strong" and "weak" forces are not forces.

Everything is consequence—the natural, inevitable outcome of ripple tension interacting with SVE gradients.

This chapter brings together all scales of interaction, from atoms to galaxies, showing that nature never needed forces at all—just one continuous field, one universal logic, and one principle: *Tension seeks equilibrium.*

11.1 — Attraction: The Relief of Tension, Not a Pull

When two opposite curvatures—like an electron and a proton—come near each other, we say *"they attract."* But the deeper truth is simpler, and far more elegant:

Opposite curvature patterns reduce total tension when they draw closer. The field, seeking relief, rearranges itself into the lower-tension state. It looks as if something is pulling—but nothing is pulling.

Here is how attraction actually unfolds:

- electrons carry a swirl with outward tension
- protons carry a swirl with inward tension
- their fields fit together like puzzle pieces
- when they overlap, the total curvature smooths
- the field prefers the smoother configuration
- ripple tension drops
- motion follows
- the electron appears to fall into orbit around the proton

This "falling" is not a force. It is simply the field descending into a lower-tension arrangement.

No force. No carrier. No exchange. Only geometry.

Attraction dissolves when we stop seeing objects and start seeing curvature. Nature rewards alignment. That reward looks like motion. We call it attraction.

11.2 — Repulsion: The Avoidance of Excess Tension

Two electrons never "push" each other apart. There is no tiny agent transmitting a shove. There is no messenger photon. There is no force field exerting influence.

What happens instead?

When like curvatures come too close, the local field tension spikes beyond what the SVE can sustain. So the field relaxes by moving them apart.

Repulsion is not a push—it is an escape. The field is simply choosing the configuration with lower strain.

Consider two inflated balloons pressed together. They seem to repel, but no force is exchanged. It is simply the geometry of pressure.

Likewise:

- like charges repel
- parallel currents repel
- negative ions repel
- even social systems behave this way metaphorically

Repulsion is just the universe saying: *"This arrangement hurts. Let's stop doing this."*

It is the alleviation of tension, not the exertion of force.

11.3 — Gravity: The Smoothest Path in a Curved Field

Gravity is where the illusion of force was most deeply ingrained into science. From Newton's apple to Einstein's curvature of spacetime to modern gravitational waves—every model tried to explain attraction using forces or geometry.

But gravity becomes trivial when we adopt SVE + RFD thinking: massive ripple structures create dense curvature in the field. Other ripples simply follow the gradient that minimizes tension.

Nothing pulls them. Nothing acts across distance. No force carrier travels between bodies.

Gravity is the universal drift toward lower-tension zones.

Let's break it down with clarity:

- A mass is a dense ripple architecture.

- Dense curvature means local SVE compression.
- Compression means a gradient in ripple tension.
- Other ripples—whether electrons, atoms, or planets—naturally move along these gradients.
- That motion looks like attraction.
- But it is guided by geometry, not force.

The apple falls because:

- the field around Earth is denser near the surface
- the field prefers the arrangement where the apple sits lower
- the apple's ripple structure moves toward that minimum
- the transition appears as acceleration

The apple never "feels" a pull. It simply follows the path of least field tension.

Gravity is alignment. Gravity is gradient flow. Gravity is the universe seeking equilibrium.

Not a force. Not a pull. Not a mystical curvature. Just consequence.

11.4 — Why Orbits Exist: The Balance of Tension and Motion

Why doesn't the Earth fall into the Sun? Or the Moon fall into Earth? Or electrons collapse into nuclei?

Because the field offers not just a downhill slope but a curved path—one where inward tension is balanced by sideways motion.

Orbits are stable paths where the tendency to minimize tension matches the ripple's momentum.

Here is how it works:

- The Sun creates a massive SVE gradient.
- The Earth moves along that slope.
- But Earth also carries sideways motion (momentum).

- The two combine into a loop around the Sun.
- The field tension is minimized in orbit, not in collision.

This is why:

- orbits are elliptical
- satellites stay aloft
- binary stars rotate gracefully
- galaxies swirl in grand spirals

Orbits are not held together by gravity. Orbits are the field's preferred compromise between tension and inertia.

Again, no force—just optimization.

11.5 — Bonding: Why Atoms Join Without Being Pulled Together

Chemical bonds—ionic, covalent, metallic—seem like forces acting between atoms. But in the field view: bonds form when electron ripple-knot architectures align in a shared, lower-tension configuration.

There is no "attractive force" in the traditional sense. There are only configurations that reduce the strain in the field.

Consider:

- **Covalent bonds** happen when electrons can occupy a mutual alignment that lowers total tension for all atoms involved.
- **Ionic bonds** happen when asymmetrical curvature distributes more evenly after an electron transfer.
- **Metallic bonds** happen when electron ripple-density becomes uniform across many nuclei.

Bonding is the universe telling us: *"This arrangement costs less tension. Let's keep it."*

No force. No pull. Only field logic.

11.6 — Repulsion Shapes the Structure of Matter

If attraction builds matter, repulsion shapes it.

- Electrons avoid collapsing into nuclei because that would spike local tension.
- Inner-shell electrons create exclusion zones because their curvature patterns overlap too densely.
- Pauli exclusion emerges not from quantum mysticism but from the field's inability to support identical overlapping curvature patterns.
- Crystals grow because certain arrangements minimize combined tension.
- Matter occupies space because curvature knots cannot overlap without destabilizing.

Repulsion is the geometry of "too much tension," not an active push.

It is the universe saying: *"This space is already occupied."*

11.7 — Gravity, Charge, Magnetism: One Principle, Many Expressions

The beauty of the field model is its unity. All interactions—

- gravitational
- electric
- magnetic
- chemical
- nuclear

—arise from the same source: *ripple tension in SVE gradients.*

Let's strip it down:

1. **Gravity** → tension gradients created by dense ripple architectures (like planets or stars)
2. **Electricity** → tension gradients created by swirl asymmetries of charged ripple-knots
3. **Magnetism** → tension gradients created by aligned internal swirls of spin structures
4. **Nuclear interactions** → tension gradients inside composite ripple architectures (protons, neutrons)
5. **Chemical bonding** → tension relief from shared electron curvature

Everything is the field seeking stability—not forces competing for dominance.

The universe does not have four fundamental forces. It has one fundamental tendency: *Minimize tension, maximize coherence.*

Everything else is detail.

11.8 — Visualizing All Interactions as Field Behavior

To grasp the unity of this idea, imagine a flexible sheet stretched across a table:

- Place a heavy ball → the sheet curves (gravity).
- Drop a marble → it rolls toward the curve (attraction).
- Add two marbles close → their depressions distort and repel.
- Spin a marble quickly → its motion balances the inward slope (orbit).
- Arrange marbles symmetrically → a stable cluster forms (bond).

Everything emerges from the sheet's geometry—no extra forces required.

The SVE field behaves the same way, but in three dimensions and with ripple curvature rather than physical dents.

Interactions are patterns. Forces are illusions.

11.9 — What Physics Gains by Abandoning Force

When force dies, clarity is born. Ending the force concept:

- unifies quantum and classical physics
- dissolves paradoxes between GR and QFT
- eliminates the need for imaginary carriers
- explains coherence across scales
- simplifies equations
- offers deeper predictive power
- aligns with observation, not metaphor
- clears space for true unification

Most importantly: it lets the universe be itself—simple, continuous, elegant.

With force gone, and interactions reinterpreted as natural consequences of ripple tension and SVE gradients, we are now ready to step into the domain where all these insights converge:

- cosmic behavior
- particle stability
- energy flows
- field architecture
- the underlying mechanics of the universe

The next chapters show how this unified field behavior explains not only motion and interaction but also the architecture of the cosmos itself.

We now leave the idea of forces behind and enter the pure logic of the field.

Interlude — When the Universe No Longer Needs Forces

A reflection at the end of Part III

There is a strange kind of silence that follows a profound realization—the silence that comes when an idea you carried your whole life suddenly dissolves in front of you.

That is the silence we have earned at the end of Part III.

For centuries, force was our anchor. It was the language through which we explained gravity, motion, attraction, repulsion, bonding, decay, orbits, tides, sparks, and stars. Force was the handle we grabbed whenever reality felt too vast to grasp directly.

But the universe was never using it. We were.

And now, having seen the field clearly—its ripples, its gradients, its perpetual search for equilibrium—we understand why force always felt a little mysterious, a little metaphoric, a little incomplete.

Because it was.

Force was a story we told ourselves to make a continuous universe feel like a set of discrete events.

Part III showed us that the story is no longer needed.

We saw attraction lose its pull and reveal itself as tension relief. We saw repulsion shed its push and emerge as overlap avoidance. We saw gravity drop its mystical authority and return to simplicity—a gradient field guiding ripple architectures. We saw bonding, chemical orbits, magnetic behavior, galactic rotation, planetary paths—all emerge from the same fundamental logic.

Nothing pushed. Nothing pulled. Nothing acted through intermediaries. Nothing exerted anything on anything else.

The universe behaved because it had to—because the field preferred certain configurations over others, because tension will always flow downhill, because gradients will always guide motion, because coherence will always win over chaos.

Part III closed the age of forces. And with it, the last major pillar of fragmented physics began to fall.

Now something remarkable happens. When forces disappear, so do the problems that forces were invented to hide.

Quantum paradoxes no longer look paradoxical. Cosmological mysteries lose their exotic drama. Black holes stop being monsters of infinite extremes. The conflict between general relativity and quantum mechanics—the "impossible unification"—reveals itself as a misunderstanding born from two incomplete interpretations.

The noise softens. The contradictions fade. The puzzles become patterns.

What emerges is a universe that is continuous, coherent, predictable in its logic, simple in its architecture, and astonishingly elegant.

Part III has cleared the ground. Part IV now builds the correction.

This next part is where physics heals—where dualities dissolve, paradoxes unclench, and all the patched-up inventions carried for decades finally lose their relevance.

We are about to reinterpret:

- collapse without mysticism
- entanglement without magic
- uncertainty without metaphysics
- dark matter without ghosts

- dark energy without fantasies
- black holes without singular mythology
- and the GR–QM conflict without impossible bridges

Part IV is not destruction. It is liberation—the release of physics from the conceptual cages it built for itself in the absence of a unified field logic.

Ahead lies the part of this book that will make many readers pause, smile, argue, and then realize—

"It always had to be this way."

Welcome to *Part IV — Dissolving the Patches & Paradoxes.*

Part IV — Dissolving The Patches & Paradoxes

"Paradoxes are not flaws in nature—they are fractures in our interpretation."

Short Overview of Part IV

Part IV is the moment where the fog finally lifts.

Everything we once labeled mysterious, paradoxical, or exotic in physics was never a mystery of the universe—it was a symptom of misreading a continuous field through a fragmented lens.

This part exposes how many of the deepest puzzles in modern science were self-created illusions: quantum duality that never existed, wavefunction collapse that was never a collapse, entanglement that was never spooky, uncertainty that was never metaphysical, dark matter that was never matter, dark energy that was never energy, singularities that were never infinite, black-hole information loss that was never loss, and the GR–QM conflict that was never a fundamental contradiction

These puzzles survived for decades because we forced the universe to behave like a set of particles and forces—a worldview the universe itself never cooperated with.

Part IV ends that struggle.

Building on the logic of SVE and ripple dynamics, each chapter reinterprets a major "mystery" as nothing more than field behavior misunderstood.

- Quantum paradoxes dissolve when wave-particle duality is replaced with ripple architecture.
- Cosmology stabilizes once we stop inventing invisible substances to rescue broken equations.
- Black holes become comprehensible when we understand ripple-density collapse without invoking infinities or magical horizons.
- And the century-old conflict between relativity and quantum mechanics ceases to exist entirely when forces dissolve and fields become the only language.

Part IV is the repair manual for the universe as we know it—a step-by-step removal of conceptual scaffolding that remained long after the building was constructed.

By the end of this part, the reader will see that the universe never contradicted itself. It only contradicted our assumptions.

What awaits is clarity—simple, continuous, unified clarity—the kind of clarity physics has been moving toward for a hundred years without realizing it was only one paradigm shift away.

Welcome to the part where problems disappear and understanding finally arrives.

Chapter 12 — Ending Quantum Paradoxes

Collapse, duality, entanglement, uncertainty reinterpreted

If you talk to students, researchers, or even seasoned physicists, you'll hear the same confession whispered with equal parts awe and frustration: *"Quantum mechanics works... but no one understands it."*

For a century, this idea became a badge of honor—as if not understanding reality was itself a mark of sophistication.

But what if quantum mechanics isn't strange at all? What if the only strange thing was our attempt to force continuity into particles, and simplicity into abstractions that were never meant to be literal?

What if the quantum "weirdness" was born not from nature, but from our metaphors?

Wave–particle duality, collapse, entanglement, uncertainty—these were not features of the universe. They were warning signs. Cracks. Signals that we were forcing nature into a frame it never fit.

Once you replace particles with ripple dynamics, and forces with field consequence, the paradoxes evaporate like mist under the morning sun.

This chapter shows how quantum mechanics becomes the simplest, most intuitive behavior imaginable when viewed from the continuous SVE field.

Let's take each paradox one by one—and watch it dissolve.

12.1 — Wave–Particle Duality: A Case of Mistaken Identity

Duality is the first quantum story every student hears—and the first misunderstanding they absorb.

A photon "acts like a wave" in one experiment, and "acts like a particle" in another. An electron "interferes with itself" as if it were a wave, yet leaves discrete dots on a detector as if it were a particle.

But here is the real story: nothing in the universe behaves like a particle. We invented that idea because we saw discrete detection events.

Every ripple-knot has two aspects:

1. It is extended → because it is a ripple in a field
2. It is detected discretely → because collapse happens locally

The duality was not in nature. It was in our interpretation.

An electron interferes because it is a ripple. It hits a screen discretely because the ripple architecture collapses locally. A photon spreads through slits as a field disturbance and deposits energy in one grain of photographic film.

Wave–particle duality = continuous ripple behavior + localized interaction events misinterpreted as "particle nature."

The mystery dissolves when we stop demanding discreteness from nature and start respecting field continuity.

12.2 — The Collapse That Never Collapsed

Quantum collapse is the most theatrical concept in physics—the idea that the entire universe waits for an observer to make up its mind.

But collapse is simply: a ripple pattern reorganizing into the lowest-tension configuration once it interacts with a dense ripple architecture (detector).

There is no metaphysical jump. There is no magical wavefunction discontinuity. There is no observer-demanded drama.

When a ripple (electron, photon, neutrino) interacts with a dense structure (atom, detector), the SVE field chooses a configuration that minimizes local curvature tension. The ripple architecture "locks" in one spot, and the interaction completes.

We call this "collapse." But it is simply:

- a reconfiguration of field curvature
- triggered by interaction
- consistent with ripple grammar
- inevitable given the gradients involved

No magic. No observer dependency. No metaphysical role for consciousness.

Collapse is not a quantum phenomenon. Collapse is structural settling.

12.3 — Entanglement: A Shared Ripple, Not Spooky Action

Einstein famously called entanglement *"spooky action at a distance."* But the only spooky part was our assumption that entangled particles were separate things.

In UED/RFD, entanglement is profoundly simple: two so-called "particles" are not two objects. They are one extended ripple configuration with two points of expression.

When you measure one point, you are not sending information or triggering action. You are simply completing a configuration that already exists across the field.

Imagine a guitar string. Pluck it in one place, and the entire string changes vibration. Not spooky. Just connected.

Entanglement is the same:

- one ripple
- shared curvature
- distributed through space
- expressed at multiple locations

There is no information transfer. No faster-than-light signaling. No metaphysical bonding.

Just one field with shared structure.

Entanglement only seemed magical because we assumed "two particles" where there was only one ripple.

12.4 — Uncertainty: Not Mystery, Just Ripple Nature

Heisenberg's uncertainty principle became a philosophical volcano. Students were told: *"You can't know both position and momentum." "Nature is fundamentally indeterminate." "Information has limits." "Reality is fuzzy."*

But all of this was misdirection.

Here is the actual reason behind uncertainty: a ripple cannot have perfect localization and perfect motion coherence at the same time, because these correspond to incompatible curvature patterns.

- Position = tight curvature localization
- Momentum = smooth, extended ripple alignment

You cannot fit both patterns into one ripple simultaneously without violating the field's geometry.

Uncertainty is not ignorance. It is not indeterminacy. It is not nature hiding information.

Uncertainty is simply the geometry of ripples.

Try to compress a water wave into a tiny region: it loses coherent flow. Let it flow smoothly: it spreads out.

The same is true for electrons, photons, all field ripples.

Uncertainty is not a paradox. It is the signature of continuity.

12.5 — Why Quantum Mechanics Looked Weird to Us

Quantum mechanics became strange because we tried to reconcile:

- the mathematics of waves with
- the intuition of particles.

The contradictions came from us, not from nature.

We demanded:

- path + position
- local + extended
- discrete + continuous
- separate + deeply connected
- instantaneous + limited by speed
- deterministic + probabilistic

These contradictory expectations were the origin of quantum weirdness.

Once the particle myth dissolves, so does the weirdness.

Electrons never walked through "both slits." Waves never collapsed magically. Entanglement was never action at a distance. Uncertainty was never mystical. Quantum randomness was never metaphysical.

All of these illusions arose from forcing particle metaphors onto a field universe.

The quantum world was never strange—only our language was.

12.6 — The Continuous Universe Behind Every Quantum Act

After dissolving these paradoxes, the quantum world becomes breathtakingly simple:

- one field
- continuous structure
- ripple dynamics
- gradient interactions
- tension minimization
- alignment optimization

Nothing is discrete. Nothing jumps. Nothing is dual. Nothing collapses mysteriously. Nothing breaks causality. Nothing violates reality.

Quantum mechanics works not because it is mysterious, but because it accidentally captured the mathematics of fields in motion—without understanding their true origin.

Once we add the missing foundation—the structured vacuum field—the entire quantum world becomes intuitive.

The paradoxes were never signs of complexity. They were signs of a wrong model.

With quantum paradoxes dissolved, we now face the next frontier of confusion: cosmology.

Dark matter. Dark energy. Inflation. Singularities.

All invented to patch equations built on flawed assumptions.

In the next chapter, we reinterpret the universe's largest mysteries using the same continuous field logic—and watch them fall just as cleanly.

Up next: "Chapter 13 — Fixing Cosmology Without Exotic Inventions."

Chapter 13 — Fixing Cosmology Without Exotic Inventions

Dark matter, dark energy, singularities, inflation demystified

Modern cosmology often reads like a dramatic script. We are told that: 95% of the universe is "missing," gravity behaves "incorrectly" inside galaxies, expansion is "mysteriously accelerating," the early universe "inflated faster than physics allows," and black holes conceal "infinite densities" at their cores.

Whenever reality refused to fit the equations, physics reached for an invisible patch—dark matter, dark energy, inflation fields, singularities. It became a habit: if the data resisted the theory, invent something that could never be directly detected.

The result was a cosmos filled with placeholders:

- matter we cannot see,
- energy we cannot measure,
- fields we cannot verify,
- infinities we cannot interpret.

The universe began to resemble a crime scene where the evidence was missing, and investigators invented suspects simply to keep the narrative alive.

But what if the universe was never missing anything? What if the patches were only necessary because our model of the field was incomplete? What if all those "mysterious substances" were nothing more than the consequences of misunderstanding the SVE field?

This chapter rebuilds cosmology exactly as the universe behaves—continuous, structured, and governed by gradients, tensions, and ripple architectures.

Viewed through this lens, the four great "mysteries of modern cosmology" dissolve effortlessly.

It is time to fix cosmology—without adding anything new at all.

13.1 — Dark Matter: The Ghost That Never Existed

Dark matter was proposed because galaxies rotate "too fast" for the visible mass they contain. Stars on the outer edges do not slow down the way Newtonian physics expects.

To fix this discrepancy, cosmologists invented an invisible, undetectable substance that could provide extra gravitational mass.

But the real issue was never missing matter. It was missing field logic.

The UED/RFD Explanation: *ripple-density gradients inside galaxies modify the effective gravitational landscape.*

Galaxies are not governed by Newtonian gravity acting on isolated particles. They are massive, spinning ripple structures embedded in a curved SVE field.

Key insights:

- Dense central regions compress the field.
- This compression propagates outward as a gradient.
- Rotational motion increases lateral ripple tension.
- This tension stabilizes outer stars without needing extra matter.
- The field's curvature alone can flatten rotation curves.

In simple terms: galactic rotation is a field phenomenon, not a mass-counting problem.

You don't need invisible matter. You only need a continuous field with ripple-based tension distribution.

Dark matter disappears—not because it was disproven, but because it was never needed.

13.2 — Dark Energy: The Universe Is Not Accelerating—Its Field Is Relaxing

Dark energy was proposed because distant galaxies appear to be receding faster than expected. The interpretation was:

- expansion is accelerating
- something must be pushing galaxies apart
- that something is "dark energy"

But acceleration is only acceleration if you assume expansion should be linear.

The SVE field tells a different story.

The UED/RFD Explanation: *the universe expands because field tension relaxes after the massive early compression. This relaxation appears like acceleration.*

Key insights:

- The early universe was extremely dense (high ripple tension).
- As the field expanded and cooled, tension dropped.
- Lower tension means lower resistance to expansion.
- So expansion speeds up—not because of a force, but because the field becomes easier to stretch.
- This natural relaxation curve mimics "acceleration."

There is no mysterious energy pushing galaxies apart. The universe is not fueled by a hidden substance.

It is simply stretching into ease—like a rubber band that relaxes when released.

Dark energy dissolves because expansion is a field consequence, not an external force.

13.3 — Singularities: The Myth of the Infinite Point

The word *singularity* entered physics because equations broke down at high curvature. Black holes, the Big Bang—both were interpreted as infinite-density points because mathematics produced infinities when pushed too far.

But an infinity in mathematics is not a physical reality. It is a signal that the model has exceeded its domain.

In the SVE perspective, there are no infinities. There is only maximum ripple-density—the highest curvature the field can sustain before reconfiguring.

The UED/RFD Explanation: *singularities are limit points where ripple-density saturates and the field reorganizes, not infinite collapses.*

Key insights:

- The SVE field has a maximum density threshold.
- When curvature approaches this limit, the field stops compressing further.
- It transitions into a stable configuration (event horizon, dense core).
- There is no infinite density, no infinite gravity, no mathematical discontinuity.

Imagine compressing a spring: beyond a point, it resists further compression and pushes back with full force. Black holes do the same on the cosmological scale.

Singularities disappear because the field is never infinite—only maximally dense.

13.4 — Inflation: A Patch for an Imaginary Problem

The Big Bang model appeared to have problems:

- Why is the universe so uniform?
- Why is the cosmic microwave background so smooth?
- Why is space flat on large scales?

To fix these issues, cosmologists invented inflation—a hypothetical period of faster-than-light expansion immediately after the Big Bang.

But inflation was a band-aid, not an explanation. It patched the symptoms of assuming that the early universe behaved like particles in a container.

The SVE field offers clarity:

The UED/RFD Explanation: *the early field was superdense and tightly coherent, so its expansion naturally preserved uniformity without needing inflation.*

Key insights:

- In a hyper-dense field, ripple architecture is globally correlated.
- Early expansion maintains coherence because ripples have not yet decohered into complexity.
- The field stretches uniformly due to its symmetry.
- No additional super-expansion is required.

Inflation was only necessary because of a particle-centric interpretation. Once we acknowledge that the early universe was a field, not a soup of particles, the "problems" inflation was meant to solve simply never existed.

Inflation dissolves because the field itself explains uniformity.

13.5 — The Universe Becomes Simple Again

For the first time in decades, cosmology becomes intuitive. There is no missing mass, no missing energy, no infinite points, no impossible speeds, no magic expansions, no invisible particles, and no exotic substances.

What remains is simply field gradients, ripple density, tension distribution, symmetry constraints, natural expansion, coherent evolution, and curvature logic.

Nothing is added, nothing invented, nothing forced. The "dark" universe disappears not because we discovered new particles, but because we discovered a better paradigm.

13.6 — What Cosmology Gains Without Exotic Inventions

By abandoning the patches, cosmology finally gains what it has needed for decades.

It achieves conceptual unity, predictive clarity, cross-scale coherence, and intuitive behavior.

The contradictions between general relativity and quantum mechanics vanish, hidden substances are no longer required, infinities dissolve, and mathematical emergencies cease to exist. Most importantly, the universe begins to look like one continuous system rather than a pile of disconnected mysteries.

The SVE field emerges as the backbone that makes cosmology whole.

With the largest cosmological mysteries resolved, we can now turn to the most mythologized object in physics—the black hole. It is not a monster of infinities, but a simple field configuration in extreme ripple-density.

Up next: "Chapter 14 — The Black Hole Without Mythology."

Chapter 14 — The Black Hole Without Mythology

Ripple-density collapse, information preservation, SVE equilibrium

If you open any popular book on black holes, you'll meet a creature straight out of myth: a place of infinite density, infinite curvature, infinite gravity, a point where physics supposedly breaks down, and a boundary called the event horizon where reality itself is said to stop cooperating. Black holes became the dragons of modern astrophysics—terrifying, mysterious, and almost supernatural.

Not because the universe behaves that way, but because our equations—built on incomplete assumptions—ran out of breath at extreme conditions. When the math diverged, we did not question the math. We mythologized the result.

But in a continuous SVE field, nothing is infinite, nothing collapses to a point, nothing breaks, nothing disappears, and nothing magical happens at the horizon. Black holes become exactly what they should have been from the beginning: extreme ripple-density architectures where the field reaches its maximum allowable curvature and then stabilizes into equilibrium.

There is no magic, no singularity, no information loss, and no event-horizon metaphysics. There is only the field behaving consistently, even at the highest densities.

This chapter strips away the myths, the exaggerations, and the mathematical ghosts—revealing black holes as elegant, predictable, stable field configurations.

14.1 — Collapse Is Real, But the Singularity Never Happens

In mainstream physics, gravitational collapse is often described as matter falling inward until its volume shrinks to zero and its density becomes infinite. But infinity is not a physical state—it is an error message.

In a continuous SVE field, ripple-density has an upper limit beyond which the field cannot compress further. The SVE behaves like a flexible but finite fabric: it can be stretched, it can be compressed, but beyond a certain threshold it becomes incompressible and redistributes tension instead of collapsing.

Black hole formation, seen through this lens, unfolds naturally. A massive star collapses, field curvature increases drastically, ripple-density approaches saturation, and the field resists further compression. Collapse halts at a stable configuration, and a dense, non-singular core emerges.

The so-called "singularity" never forms because continuous curvature cannot be compressed beyond the field's intrinsic limit. Instead, the field redistributes tension rather than allowing infinity, and the collapse stabilizes into an equilibrium that is finite, consistent, and smooth.

The black hole's core is not a point. It is a maximally compressed but continuous region—a knot of extreme curvature, not an impossible singularity. The monster evaporates, and the physics becomes simple.

14.2 — The Event Horizon: Not a Wall, Not a One-Way Gate

The event horizon is usually described as a boundary of no return, a place where time stops, a region where light cannot escape, and a surface where physics behaves differently. Yet the SVE interpretation is far more grounded. The event horizon is simply the radius where outward-propagating ripples can no longer overcome the inward field gradient.

There are no walls, no magic, no frozen time, and no metaphysics.

Think instead of a strong river current. If you drop a leaf upstream, it can float away freely. But if you drop it too close to the waterfall, the current pulls it inward faster than it can drift outward. A black hole works in precisely the same way: ripple gradients flow inward, outward propagation becomes impossible, and signals are trapped because the slope is too steep.

Nothing special happens at the horizon. It is not a physical membrane but a geometric threshold, a point in the field gradient where outward motion is no longer sustainable. The event horizon only looks dramatic if you imagine space as empty and gravity as a force. Within a continuous SVE, it is simply the steepest part of the field.

14.3 — Information Never Dies, Because Ripples Never Stop Existing

One of the biggest modern puzzles—the so-called *black hole information paradox*—arose from a chain of assumptions: matter was imagined as particles, particles were thought to carry information, particles were believed to fall into a singularity, singularities were assumed to destroy information, and therefore information was said to vanish.

But if you replace particles with ripples, the paradox dissolves instantly.

What really happens is straightforward. Matter falling into a black hole becomes increasingly compressed ripple structures. Their ripple identities do not vanish; they reorganize within the dense field. The black hole's internal architecture encodes every incoming curvature signature, while Hawking-like radiation transmits structural reconfigurations outward over immense timescales. The universe never loses information—it only redistributes it.

Information is not a property of particles. It is a property of field curvature patterns. And curvature cannot disappear. It can only stretch, reconfigure, compress, radiate, or equilibrate.

Black holes preserve information because the field itself preserves the continuity of curvature. The paradox never existed. Only the particle myth did.

14.4 — The Black Hole Interior: A Continuous, Stable Ripple-Knot

Imagine the black hole interior not as a point of collapse, but as the densest ripple architecture in the universe—a region where curvature saturates, a domain of maximum SVE compression, and a stable equilibrium rather than a breakdown.

This interior behaves as a continuous structure, not a singular point. It is a finite, dense core, not an infinity. It is a stable configuration, not a violation of physics.

Inside a black hole, time does not stop, space does not disappear, physics does not break, and ripples do not vanish. Instead, the field enters a regime where its behavior is dominated by extreme gradients, yet remains fully governed by the same rules that shape electrons, stars, and galaxies.

A black hole interior is simply the high-density limit of continuous field logic. Nothing magical occurs, nothing violates natural law.

14.5 — Hawking Radiation Without Mysticism

Hawking radiation is traditionally explained through quantum vacuum fluctuations and particle–antiparticle pair creation at the event horizon. But if particles do not exist, this interpretation collapses.

The field explanation is far smoother. Extreme curvature at the horizon forces a reconfiguration of local ripple patterns, and these reconfigurations propagate outward as faint radiation. There are no virtual particles, no negative-energy ghosts, and no spontaneous creation or annihilation. What occurs instead is curvature redistribution, ripple alignment shifts, tension adjustments, and outward-propagating field disturbances.

Black holes "evaporate" because the field slowly smooths its extreme curvature. Hawking radiation is nothing mystical—it is simply the field exhaling tension.

14.6 — What Happens at the Core? Equilibrium, Not Infinity

At the center of the black hole lies not a singularity but a ripple-density equilibrium zone. In this region, curvature stops increasing, field resistance reaches its maximum, compression saturates, tension redistributes smoothly, and structure stabilizes.

This zone is the stability floor of the SVE field. Just as electrons have stability limits that prevent collapse, black holes have their own limits. The field does not break—it stabilizes.

Nature hates infinities. It never creates them.

14.7 — The Elegant Simplicity of the Field-Based Black Hole

With the SVE lens, a black hole is revealed not as a hole, not as a singularity, not as a tearing of spacetime, and not as a one-way cosmic portal. It is instead a maximum-density ripple configuration—stable, continuous, finite, information-preserving, and perfectly lawful.

The universe never violates itself. We only violated interpretation.

With black holes demystified and singularities removed from physics, only one major conceptual knot remains: the alleged conflict between General Relativity and Quantum Mechanics. Yet as we will see, the conflict was never real. It arose only from our attempt to reconcile two incomplete pictures.

Up next: "Chapter 15 — The GR–QM Conflict: A Problem That Never Needed Solving."

Chapter 15 — The GR–QM Conflict

A Problem That Never Needed Solving

Unifying everything through continuous field logic

For a century, physics has lived under a peculiar psychological burden. On one side stood Einstein's General Relativity—smooth, majestic, geometric—describing the universe as a fabric of curvature. On the other stood Quantum Mechanics—raw, statistical, jittering—filled with probabilities, jumps, and paradoxes. Together, they created the narrative of a "broken universe," two masterpieces that refused to merge.

The contradiction was considered so deep that entire careers, institutes, and industries were dedicated to "solving" it. String theory, loop gravity, extra dimensions, quantum foam, gravitons, holographic universes—all were attempts to force two mismatched worldviews to coexist.

But what if the mismatch was never in the universe, but in the assumptions behind those theories? What if General Relativity and Quantum Mechanics only conflict because both were built on incomplete foundations? What if unification does not require new dimensions, exotic strings, or cosmic holograms, but simply a return to continuity?

This chapter makes a bold claim: the GR–QM conflict is an illusion. Both are incomplete approximations of the same continuous

SVE field. Once particles, forces, and discrete assumptions dissolve, the conflict evaporates. There was never a war between theories. There was only a misunderstanding between interpretation and reality.

15.1 — Why General Relativity Works (But Not Fully)

General Relativity is built on a beautiful idea: mass curves spacetime, and everything follows that curvature. This vision has proven astonishingly successful. It explains the motion of planets and stars, the bending of light around massive bodies, the horizons of black holes, the rippling of gravitational waves, and even the grand expansion of the cosmos itself.

Yet beneath its elegance lies a set of hidden assumptions. General Relativity imagines the universe as a perfectly smooth continuum. It allows no discreteness, no particles, no sudden jumps, no forces, and no fields separate from geometry. Reality, in this picture, is fluid curvature alone. This comes close to the SVE perspective, but it is not identical.

The cracks appear when we look more closely. General Relativity cannot handle microstructure. It collapses into singularities when pushed to extremes. It relies on stress-energy tensors that fail to capture the true dynamics of ripples. It treats curvature as abstract geometry rather than as a physical field with tension limits. And it offers no mechanism for ripple stability at the smallest scales.

General Relativity remains a brilliant macro-theory, a sweeping continuous picture of the universe. But it is incomplete. It describes the shape of reality without describing the material of that shape.

15.2 — Why Quantum Mechanics Works (But Not Fully)

Quantum Mechanics presents almost the opposite worldview to General Relativity. It speaks in the language of discrete measurements, quantized energy levels, wavefunctions, probabilities, entanglement, non-local correlations, and uncertainty. At its core, it assumes the presence of ripples, but interprets them only as probabilities of particle behavior. In doing so, it captures the activity of reality, yet misreads the true nature of that activity.

The shortcomings of this framework become clear once examined closely. Quantum Mechanics assumes particles exist, even though its own mathematics points to waves. It inserts collapse as a magical rule to explain measurement. It treats continuity as probability waves rather than as physical structures. It introduces duality because detection events are misinterpreted. It lacks a genuine physical substrate, offering only mathematical rules without material grounding. And when extended to cosmic scales, it fails entirely.

Quantum Mechanics, then, describes how the field behaves, but not what the field actually is. It provides a powerful map of phenomena, but the terrain beneath that map remains misunderstood.

15.3 — The Illusion of Conflict: Two Halves of One Reality

General Relativity and Quantum Mechanics only appear incompatible because they begin from opposite assumptions. General Relativity imagines a universe with no discreteness at all, while Quantum Mechanics insists that discreteness is fundamental. These two premises cannot coexist, and so the tension between them was born.

But the SVE field shows that discreteness never existed. Everything is continuous ripples, and what we call "detection events" merely create the illusion of discreteness. Once the particle myth is removed, the supposed conflict dissolves. Quantum Mechanics becomes a theory of ripple behavior, while General Relativity becomes a theory of ripple gradients. They are not different; they are complementary.

Quantum Mechanics describes local ripple dynamics. General Relativity describes large-scale ripple gradients. And the SVE field is the substrate that connects them both.

The conflict vanishes because each theory collapses into a simpler truth: geometry is ripple dynamics, and ripple dynamics are geometry. The universe is not divided between smoothness and discontinuity, between GR and QM. It is fully smooth, always. The discontinuity was never in existence itself—it was only in our interpretation of detection.

15.4 — How SVE Unifies GR and QM Effortlessly

Structured Vacuum Energy (SVE) is the missing foundation that allows General Relativity and Quantum Mechanics to merge without struggle. It carries the continuity of General Relativity, for in the SVE field there is no discreteness, no particles, no jumps, no singularities, and no broken spacetime—only one seamless field. At the same time, it carries the dynamism of Quantum Mechanics, where ripples propagate, modes interfere, and curvature patterns quantize naturally because stability favors certain configurations.

SVE also removes the need for forces in both theories. There are no carriers as imagined in Quantum Mechanics, and no action at a distance as described in General Relativity. Motion arises simply from ripple tension interacting with field gradients.

In doing so, SVE eliminates the paradoxes that have haunted physics for decades. Duality dissolves, collapse is no longer required, entanglement loses its magic, infinite density disappears, dark energy is no longer invented, and the supposed contradiction between gravity and quantum mechanics evaporates. Everything reduces to ripple configuration, ripple stability, ripple gradients, and ripple consequences.

Most importantly, SVE explains both the micro and the macro with the same rule. Atoms stabilize through field tension. Planets orbit through field gradients. Photons propagate through alignment. Black holes form through ripple-density saturation. Nothing requires separate laws. One substrate, one vocabulary, one logic.

15.5 — Why the Conflict Was a Problem of Language, Not Physics

General Relativity speaks in the language of geometry. Quantum Mechanics speaks in the language of algebra. But the universe itself speaks in the language of ripples. Both theories were shadows of the same underlying truth—projections of one reality seen from different angles.

The conflict was never physical; it was linguistic. General Relativity says that mass curves spacetime. Quantum Mechanics says that particles have quantized behavior. Yet the SVE field reveals that mass is ripple curvature, quantization is ripple stability, spacetime is the field itself, probability is unresolved ripple configuration, gravity is gradient flow, and interaction is tension minimization. Everything reduces to the same root.

For a hundred years, physics asked the wrong question: *"How do we merge two incompatible theories?"* The real question should have been: *"What foundational assumption is wrong in both?"* The answer is simple—particles and forces. Remove them, and the world becomes one story again.

15.6 — The Real Unified Theory: The Field Is the Mechanism

Unification was never about inventing new layers of complexity. It was always about removing the unnecessary ones. In the UED/RFD worldview, the universe is not a patchwork of particles, forces, and separate laws—it is one continuous field. That field carries a single grammar, a single kind of structure, a single form of interaction, a single cause of motion, and a single explanation of stability. It provides one architecture across all scales, one source of quantization, one mechanism for coherence, one substrate for gravity, one origin of cosmic expansion, one template for black holes, and ultimately one rule behind everything.

Ripple tension interacting with SVE gradients governs all of physics. This is the real unification—the one that General Relativity and Quantum Mechanics never reached, not because they failed, but because they were never meant to be final. They were stepping stones toward continuity, partial glimpses of a deeper substrate.

Once the field becomes primary, there is nothing left to unify. Everything already is.

15.7 — What Physics Gains by Ending the Fake Conflict

Dissolving the divide between General Relativity and Quantum Mechanics opens the door to a new era of clarity. Cosmology becomes unified, quantum theory regains coherence, black holes are explained intuitively, and mass and energy find natural interpretations. A single mechanism governs all interactions, forces and carriers are no longer required, paradoxes resolve themselves, infinities vanish, mathematics becomes simpler, predictions deepen, coherence stretches across scales, and experiments gain unprecedented clarity.

But above all, physics regains elegance. The universe ceases to look like a contradiction wrapped in mathematics and instead reveals itself as what it always was: a single continuous field performing an infinite symphony of ripples.

With General Relativity and Quantum Mechanics finally reconciled through the simplicity of a continuous field, we now stand at the threshold of something larger. The implications reach far beyond theory. Physics education must be rewritten. Technology must be reimagined. Energy must be reconsidered. Matter must be reconceived. Cosmology must be rebuilt. And our entire scientific worldview must be restructured.

Part IV dissolved the illusions and patches. The next part will build the future.

Interlude — When Mystery Becomes Method

A reflection at the end of Part IV

We have arrived at a rare and quiet moment in the journey—the pause after the fog has lifted but before the new horizon fully reveals itself. Part IV was not merely a chapter of explanations; it was a chapter of release.

For decades, physics wrapped itself in paradoxes, patched equations with invisible inventions, and tolerated infinities as "mathematical artifacts." Each time nature contradicted our expectations, we added something new: dark matter, dark energy, inflation, singularities, virtual particles, force carriers, quantum duality, collapse postulates, renormalization tricks. Theories grew heavier, but not clearer.

Part IV revealed what happens when we strip away these inventions and return to nature with one simple assumption: the universe is a continuous field, not a collection of objects. Once the particle lens and force-based logic were removed, everything that once looked mystical became obvious. Quantum paradoxes resolved into ripple behaviors. Cosmological mysteries became field gradients. Black holes became dense stability zones. General Relativity and Quantum Mechanics became two approximations of the same substrate.

This is not the end of mystery—it is the end of confusion. And that distinction matters. Confusion paralyzes us. Mystery inspires us. Part

IV dissolved the confusion, leaving behind a universe full of possibility, not impossible puzzles.

The transition is profound. For the first time in the history of physics, we stand in a place where nature is simple, laws are unified, behavior is continuous, paradoxes are resolved, foundations are clear, and the path forward is coherent. This is the moment when physics shifts from interpreting the universe to engineering the universe.

Until now, science has been reactive—explaining, modeling, adjusting. But once the SVE field becomes the operating framework, we gain something unprecedented: predictive, constructive control over the substrate of reality itself. This is the threshold between *knowing the world* and *shaping the world.*

Ahead lies the next era of physics—not built on particles, not powered by forces, not bound by twentieth-century constraints, but driven by the logic of field engineering. What we are about to enter is not science fiction. It is inevitable science.

Part V is where physics stops being a spectator and becomes a creator. Ripple behavior becomes technology. Gradients become propulsion. Coherence becomes medicine. The vacuum becomes an energy source. Ripple logic surpasses quantum logic. Human civilization steps into a fundamentally new phase.

The universe is no longer a puzzle. It is a platform.

Welcome to Part V — The New Frontier of Physics & Technology.

Part V — The New Frontier Of Physics & Technology

"When the field becomes familiar, the future becomes engineerable." "To understand the universe is to gain access to its machinery."

Short Overview of Part V

Part V opens the door to the world that lies beyond understanding—a world where physics is no longer confined to description, and humanity begins to interact directly with the continuous field that underlies all existence.

So far, this book has dismantled the illusions of particles, forces, paradoxes, and cosmological patches. In their place we have built a unified picture: a single structured field, evolving through ripples, gradients, and equilibrium.

The natural next question arises: if the universe works through ripple tension, alignment, and field gradients, what can we do with that knowledge?

This section explores the possibilities that emerge when science stops resisting the universe's continuity and begins to use it. In this new paradigm, materials are no longer static matter but engineered ripple architectures with tunable coherence. Communication is no longer limited by photons but enhanced through field-based coupling. Computing moves beyond qubits, operating through ripple-coherence logic. Motion ceases to depend on propellant and instead rides the vacuum landscape by shaping gradients. Medicine evolves from chemistry and surgery into coherence-restoration and field stabilization. Energy systems shift from combustion and collisions to alignment-based extraction and equilibrium devices that work with the vacuum rather than against it.

The technologies of the future will not be built by pushing matter or accelerating particles, but by sculpting the field that creates matter in the first place.

Part V is not fantasy. It is the mature stage of physics—the stage where clarity allows construction, and understanding leads naturally to capability. This is the moment when humanity transitions from being a passive observer of cosmic mechanics to an active participant in shaping them.

In Part V, the universe is no longer a distant phenomenon. It becomes a platform—a dynamic, responsive field that can be directed, tuned, stabilized, amplified, and engineered. This is the new frontier of physics and technology—not built on force, matter, or particles, but on alignment, coherence, tension, and gradients within the SVE field itself.

Here begins the era where the universe's foundation becomes humanity's most powerful tool.

Chapter 16 — Field Engineering

The Next Technological Revolution

Materials, Communication, Computing, Propulsion

For centuries, technology has advanced in one direction: we push things. We push electrons through wires, fuel into engines, data across networks, rockets against gravity, and matter against matter to break it apart. Every invention we know—cars, computers, satellites, lasers—has been built on the same primitive assumption: to make something happen, you must push something else.

But what if the universe never worked that way? What if pushing was the least efficient way to interact with reality? What if the true machinery of the cosmos is not matter at all, but the field beneath it?

Imagine a different paradigm. Movement is created not by burning fuel but by shaping gradients. Computation is achieved not by flipping bits or qubits but by tuning coherence patterns. Communication occurs without transmission, through field-coupled coherence that manifests everywhere at once. Materials are designed not by mixing atoms but by sculpting ripple architectures that determine hardness, transparency, magnetism, and strength.

This is not science fiction. It is the natural next step once we understand the SVE field as the substrate of all physical behavior. For the first time in history, we are poised to stop manipulating objects and begin shaping the field that makes objects possible.

When physics becomes continuous, technology becomes limitless. Welcome to the era of Field Engineering—where the universe is no longer an obstacle but a medium, and the frontier of innovation begins with the ripples beneath reality.

16.1 — From Observing the Field to Designing It

For most of human history, physics has been a spectator sport. We observed, measured, predicted, and eventually learned to influence matter—but always indirectly, always from the outside. Engines and microchips, bridges and rockets, all emerged from our ability to push, pull, heat, cool, or break the materials in front of us.

This approach worked brilliantly as long as we believed the universe was built from objects—particles, forces, atoms—and that technology meant arranging those objects in clever ways. But in a continuous field universe, matter is not fundamental at all. It is the after-effect of ripple patterns held in stable curvature. Energy is not a commodity to be burned or generated; it is the redistribution of tension within the field. Motion is not caused by forces; it is guided by gradients that the field itself prefers.

Once this becomes clear, the old method of engineering begins to look primitive, even wasteful. We heat materials to force reactions that the field would perform elegantly if we simply tuned local coherence. We push spacecraft with tons of propellant when the field could carry them effortlessly through controlled gradient shaping. We encode information in electrons and photons when the field could store patterns directly in coherence modes.

Traditional engineering hits a ceiling because it works after the world has already formed. It manipulates consequences, never the cause. It moves matter but does not move the field beneath it.

Field engineering reverses that relationship. It is the moment when humanity stops rearranging products and begins shaping the substrate itself. The SVE field is continuous, responsive, and tunable. Every ripple can be strengthened, aligned, stretched, relaxed, or reconfigured. Every gradient can be sculpted like a landscape. Every coherence pattern can be stabilized or dissolved.

Ripple-design becomes the new architecture. Gradient-shaping becomes the new propulsion. Coherence tuning becomes the new communication, computation, and even medicine.

We move from watching the field work to working with the field. And that changes everything.

16.2 — Materials as Ripple Architectures

When we stop seeing matter as a collection of atoms and begin to see it instead as a set of stable ripple architectures within the SVE field, the entire meaning of "materials science" changes. The solid world around us—the metals we shape, the glass we mold, the semiconductors we etch—is not built from tiny objects sitting side by side. It is built from patterns of curvature, locked into place by coherence, alignment, and tension balance within the field.

Hardness is no longer the property of atoms but the measure of how strongly a ripple network resists distortion. Conductivity is not electrons flowing like marbles through pipes but a ripple pathway with vanishing tension barriers. Transparency is not empty space between atoms but a ripple architecture that allows alignment waves to pass without scattering. Magnetism is not spinning particles but a swirl-coherence maintained across a lattice of ripples. Flexibility is not softness of matter but the ease with which ripple geometry can reconfigure without losing stability.

In this view, every property of matter emerges from how the underlying field is arranged—not from what "particles" it is made of. And that means materials are no longer constrained by the periodic table or chemical bonding rules. They can be designed directly at the level of ripple behavior.

Imagine materials whose density is determined not by atomic spacing but by how closely ripple-knots are allowed to align—yielding ultra-light structural matrices stronger than steel yet almost weightless. Imagine fabrics where coherence is so perfect that resistance vanishes—true superconducting sheets operating at room temperature. Imagine metamaterials engineered not through micro-patterns but through field-level ripple modulation—producing adaptive optics, cloaking surfaces, and dynamically tunable refractive behavior. Imagine solids that respond to external gradients like living organisms to stimuli—field-responsive materials that change rigidity, shape, or conductivity based on the local SVE landscape.

This is not chemistry. It is choreography—the deliberate arrangement of ripples to produce new forms of matter that nature itself has never had the opportunity to assemble.

When matter becomes ripple architecture, material design becomes field design—and the palette of possibilities expands without limit.

16.3 — Communication & Computing Through Coherence

For more than a century, our communication systems have been built on a single assumption: to send information, you must transmit something—a photon, a pulse, a bit, a packet, a wave. Yet transmission is inherently slow, lossy, and limited. Signals weaken with distance, scatter through obstacles, and demand vast infrastructure simply to move patterns from one place to another.

The deeper problem is conceptual. Electromagnetic communication treats information as something that must travel through space. But in a continuous SVE field, nothing truly needs to travel at all—the field is already everywhere.

This is where coherence-based communication enters: a paradigm in which signals are not sent but manifested through field-coupling.

A coherent ripple pattern created in one region instantly aligns a corresponding region elsewhere, not by transmission but by synchronization within a shared substrate. This is not "faster-than-light messaging." It is no travel at all—two points resonating on the same coherent structure of the vacuum field, like two ends of a vibrating string moving together without anything passing between them.

Such coherence networks allow near-instant global coupling. Communication is no longer defined by speed, but by alignment.

Computation evolves just as dramatically. Today's processors flip bits by moving charges, while quantum computers manipulate fragile superpositions that collapse under the slightest disturbance. Ripple-based logic transcends both. Instead of switching states by moving electrons or maintaining qubits, a ripple processor shifts field configurations directly—a computation performed by rearranging curvature itself.

Coherence chains become the new logic gates: sequences of aligned ripple modes that trigger structural shifts. Stability regions act as memory: zones of locked curvature patterns that hold information without consuming energy. Gradient memory replaces magnetic domains and charge traps, storing information as differences in field tension rather than material arrangements.

Ripple-state processors emerge as the ultimate machines—devices that compute by sculpting the SVE substrate in real time, far more robust than qubits and orders of magnitude more powerful than silicon.

Communication becomes coherence. Computation becomes field choreography. And information ceases to be something we send—it becomes something we shape.

16.4 — Propulsion Through Gradient Shaping

For centuries, human movement—on land, sea, air, or space—has been chained to one primitive rule: if you want to go forward, you must throw something backward. Engines burn fuel, rockets expel mass, jets shove air, and even ion drives spit out ions. Every method depends on propellant, thrust, and friction against something external.

But this entire logic collapses once we recognize that motion is not the result of force, but the consequence of following gradients in the SVE field. If movement emerges naturally from field geometry, then propulsion becomes a matter of shaping that geometry—not fighting against gravity or pushing against air.

This is the essence of gradient propulsion: movement achieved by altering the field landscape around a vehicle so that it simply flows toward the lower-tension region. It is not thrust. It is not lift. It is not propulsion in any traditional sense. It is field-surfing—a controlled glide through the invisible currents of the vacuum itself.

Imagine creating a slight curvature asymmetry around a craft—a ripple-pressure difference ahead versus behind. The craft moves not because it pushes matter, but because the field reorganizes into a lower-tension configuration, carrying the craft along with it. Just as a leaf drifts toward the center of a whirlpool, a ship immersed in a shaped SVE gradient moves toward the direction where tension has been reduced. The ship does nothing. The field does everything.

Neutral-lift travel emerges naturally from this logic. By creating a local "buoyancy" in the field—a region where curvature beneath the craft is slightly less dense—the craft rises without wings, thrust, or combustion. It becomes the equivalent of a helium balloon floating upward, but in vacuum, without air, using only the gradient of the SVE substrate.

Local curvature shaping acts like the rudder and sail of this new propulsion paradigm. Small ripple modulations determine direction, speed, ascent, and descent. Larger gradient shifts enable long-distance travel, orbital insertion, or smooth hovering without consuming any reaction mass.

In this system, ships are not pushed—they glide, effortlessly and silently, like objects floating through a fluid that happens to be everywhere.

Propulsion becomes ripple navigation, not energy expenditure. Movement becomes field alignment, not mechanical struggle.

This is the dawn of motion without fuel—movement by shaping the universe itself.

16.5 — The Birth of a Unified Engineering Paradigm

There comes a moment in every scientific era when separate disciplines suddenly reveal themselves as fragments of one deeper truth. Field engineering is that moment—the point where the artificial walls between physics, materials science, computing, communication, propulsion, and energy quietly dissolve.

When we stop treating the universe as a warehouse of particles and begin to see it instead as a continuous, structured field, something remarkable happens: all technologies reveal themselves as variations of the same act—the act of shaping ripples.

Materials become stable ripple architectures. Energy becomes the redistribution of tension. Movement becomes gradient navigation. Communication becomes coherence alignment. Computation becomes controlled ripple reconfiguration. Five different industries collapse into one underlying mechanism, one substrate, one language.

The future laboratory will no longer be a room filled with chemicals, circuits, magnets, fuels, or mechanical systems. It will be a ripple-design environment—a place where engineers sculpt curvature, stabilize coherence, and shape gradients like artists working with light.

Instead of measuring voltages, we will measure tension flows. Instead of manipulating atoms, we will manipulate ripple signatures. Instead of forcing reactions, we will tune field alignment. Instead of building engines, we will sculpt gradients that allow matter to glide.

This shift is deeper than the move from mechanical machines to electronics, or from electronics to digital systems, or from digital systems to quantum devices. Those transitions changed how we build things. Field engineering changes what it means to build anything at all.

It is not mechanical. It is not electronic. It is not quantum. It is field-centric—continuous, coherent, and universal.

Every future device—from propulsion systems to medical instruments to energy generators to computation platforms—will operate using the same underlying logic: shape the field, and reality responds.

This is not the next technological revolution. It is the beginning of a new era—a unified engineering paradigm where humanity stops rearranging matter and starts collaborating with the universe's foundational fabric itself.

Chapter 17 — Gravity Control Through SVE Gradients

Non-propellant movement, orbital manipulation, field surfing

Every technology humanity has ever built—from bicycles to rockets—has treated gravity as an adversary. We burn fuel to lift off, consume energy to climb, and construct heavy infrastructure just to rise a few meters above the ground. Even in space, controlling orbit requires relentless thrust, constant correction, and enormous expenditure. For centuries, we assumed this was the unavoidable price of motion. Gravity was a force to resist, overcome, or escape. Movement meant pushing against something—air, ground, water, exhaust—anything that could provide leverage.

But all of this struggle was based on a misunderstanding. Gravity was never a force. It was a gradient—a gentle slope in the SVE field that objects naturally follow. And the moment we accept this, a door opens to something extraordinary.

If gravity is a gradient, then motion is not a battle—it is navigation. Imagine rising from the ground without fuel, not by pushing upward, but by shaping the field beneath you so that the path of least tension moves upward instead. Imagine a spacecraft gliding silently through vacuum, not by burning propellant, but by surfing a curvature corridor designed around it like an invisible river. Imagine altering an orbit simply by tuning the local field density until the craft drifts into a higher or lower shell without expending a single drop of fuel.

This is not antigravity. It is not exotic matter. It is not reactionless propulsion. It is gradient engineering—controlling motion by sculpting the very landscape from which gravity itself arises.

For the first time in human history, we stop fighting gravity and begin collaborating with it. Welcome to the physics of field-surfing, neutral-lift travel, and non-propellant movement—the future of motion in a continuous universe.

17.1 — Gravity as a Gradient, Not a Force

The idea that gravity "pulls" is one of the oldest intuitions in human thought. We feel weight pressing down, we watch objects fall, we witness planets orbit—and it seems natural to imagine an invisible hand tugging everything inward. Yet this picture collapses the moment we adopt the continuous SVE field as the true substrate of the universe.

In the field view, gravity is not an agent. It is not a force. It is not something that reaches out across space to drag masses toward one another. Gravity is simply the shape of the field—a slope carved into the SVE landscape by dense ripple architectures such as planets, stars, or galaxies.

A mass does not fall because it is pulled. It falls because the field beneath it offers a downhill path of lower tension. Just as a marble rolls along a curved surface without being "forced," objects move through the SVE field by following the gradients of curvature around them. The field reshapes itself to reduce tension, and anything immersed in it drifts naturally along that reshaping. Movement becomes a consequence of the field seeking equilibrium—not the result of a mysterious attraction.

This shift in interpretation changes everything. If gravity is a landscape, then controlling motion is not about resisting or overpowering it. It is about redesigning the terrain. Instead of thinking in terms of thrust or lift, imagine crafting a tiny, temporary slope beneath a vehicle—an engineered dip or tilt in the field's curvature that makes "downhill" point in the direction you want to go. No force. No propellant. Just following the new path of least tension.

In this framework, gravity is no longer an obstacle but an invitation: reshape the gradient, and you reshape motion.

This is the conceptual breakthrough of gradient physics. We stop asking, *"How do we fight gravity?"* and begin asking, *"What shape must the field take for motion to occur effortlessly?"* Once we accept gravity as a gradient, the possibility of controlling that gradient becomes the foundation of a new era in propulsion and mobility.

17.2 — Creating Motion by Designing Local Field Asymmetry

Once we recognize that gravity is nothing more than a gradient in the SVE field, a remarkable possibility emerges: movement does not require pushing anything. It only requires creating a small asymmetry—a tension difference—in the field surrounding a craft.

Imagine a boat on a river. The boat does not move by pushing the water backward; it moves because the river itself flows. If we can shape the field so that the "flow" points in our preferred direction, the craft simply rides it.

This is the essence of non-propellant motion. A region of slightly lower tension is created in front of the vehicle, and a region of slightly higher tension behind it. The SVE field, always seeking equilibrium, reorganizes itself and carries the craft into the region of lower tension—not because the craft is being pulled, but because the field itself wants to relax.

Nothing is expelled. Nothing is burned. Nothing is pushed. The craft moves because the field moves.

From this logic arises the concept of field-surfing. Instead of fighting gravity or vacuum, a vehicle shapes a gentle slope in the field and glides along it like a surfer riding a wave. The craft is not thrusting forward—it is following the new downhill path created by the engineered gradient.

Vertical motion becomes equally elegant. By reducing the ripple density beneath the craft—effectively creating a local lightening of the field—the craft ascends without wings, fans, or thrust. It rises because the region below becomes less stable in curvature, and the field redistributes by lifting the craft upward. This is neutral-lift travel: gravity does not disappear; the gradient simply reorients.

Directional control becomes a matter of curvature alignment. Small adjustments in ripple orientation around the craft tilt the field landscape left, right, up, or down, enabling precise navigation with no mechanical parts.

In this world, propulsion ceases to be an act of force and becomes an act of design. We do not push the vehicle—we sculpt the field it rests in.

Motion becomes the art of asymmetry—the gentle practice of guiding the universe so that movement happens naturally, with almost no energy consumption and no struggle against nature itself.

17.3 — Orbital Manipulation Through Gradient Sculpting

In traditional aerospace engineering, orbit control has always been a relentless battle. Satellites burn fuel to correct drift, rockets expend enormous mass to change altitude, and every maneuver carries a strict penalty: use fuel or lose control. This dependency limits mission lifespans, caps maneuverability, and turns every orbital decision into a resource gamble.

But all of this only feels necessary because we misunderstand what an orbit truly is. An orbit is not a "balance of force" between gravity pulling inward and velocity pushing outward. It is simply a stable path on the SVE landscape—a contour line etched into the curvature surface surrounding a planet. And if the landscape changes, the path naturally changes with it.

This is the heart of gradient sculpting. Instead of pushing a spacecraft into a new orbit with thrust, we modify the local ripple density or curvature gradient around the vehicle. A small reduction in field density above the craft encourages it to rise into a higher orbital shell. A slight increase below nudges it downward. The craft moves not because it exerts force, but because the field offers a new slope to follow.

No rockets. No ion engines. No fuel. Just geometry.

Transitioning between orbital shells becomes as smooth as shifting lanes on a gently curved highway. By tuning gradients around the craft, we guide it toward the next stable contour, enabling orbital changes at negligible energy cost and with exquisite precision.

Station-keeping transforms even more dramatically. Today, satellites constantly fire thrusters to resist drift caused by micro-perturbations. In a field-centric framework, hovering and parking are achieved by balancing the SVE environment around the craft so that the net gradient is flattened. When there is no slope, there is no drift. The craft does not fight to stay in place—the field holds it there.

This unlocks the possibility of continuous, silent, ultra-efficient navigation through vacuum and orbital space. Movements become fluid, not impulsive. Corrections become field adjustments, not propellant burns.

Orbital mechanics evolves from a fuel-limited discipline into a precision art of reshaping the curvature around a vehicle. In this paradigm, space becomes navigable not by force, but by field intelligence.

17.4 — The Road to True Gravity Control

The journey from understanding gravity to controlling it begins with small steps—localized curvature tweaks around a craft—but it does not end there. Once we learn to shape gradients on demand, we naturally progress toward something much larger: the ability to engineer gravitational environments themselves.

The SVE field is not static; it responds to tension, coherence, and ripple density across all scales. Local shaping gives us lift, drift, and precise navigation. Large-scale modulation gives us something even more profound—the power to design field pockets, regions where curvature is intentionally softened or intensified.

A field pocket is like a temporary valley or bubble carved into the SVE landscape. Inside it, movement becomes effortless. Outside it, the natural gravitational profile remains unchanged. Such pockets could become the invisible infrastructure of future transport networks—guiding vehicles along pre-shaped tension pathways with no engines, exhaust, or mechanical noise.

Beyond pockets lie gradient wells—engineered depressions in the field that can anchor stations, platforms, or craft without physical support. A hovering city may sound fantastical today, but in a field-driven world, it is simply a matter of shaping a stable curvature basin at a chosen altitude.

Then come curvature corridors, long, smooth channels of modulated field density spanning great distances. These corridors act like cosmic highways: vehicles entering them naturally accelerate and glide, carried by the gradual slope engineered into the corridor itself. Intercontinental travel, orbital transfers, and deep-space movement all become the same basic act—surfing through pathways carved in the vacuum.

The implications are enormous. Space travel becomes silent, continuous, and nearly energy-neutral. Planetary access no longer requires massive rockets; craft simply follow rising or descending gradients. Near-surface mobility transforms: hovering platforms replace wheels, silent aircraft replace noisy combustion engines, and movement becomes smooth, safe, and in complete harmony with the environment.

Gravity control is not a distant dream. It is the first major, inevitable application of field engineering—the moment when physics stops being observational and becomes architectural.

Humanity's next mode of transport will not be about defeating gravity. It will be about designing it.

Chapter 18 — Medicine of the Future

Healing Through Field Coherence

Ripple diagnostics, coherence therapy, energy stabilization

Modern medicine has long been built on a war mentality. We attack infections, fight cancer, suppress inflammation, block pathways, override signals, and force the body into compliance. Every treatment is framed as a struggle—the body versus disease, the drug versus the symptom, the surgeon versus the defect.

But what if illness is not an enemy? What if it is simply a distortion—a ripple pattern that has lost coherence within the body's larger field architecture? In that case, the goal is not to fight, but to re-align.

Imagine medicine based not on pills and scalpels, but on coherence tuning. Imagine diagnosing disease not by the damage it has already caused, but by the subtle disturbances in the SVE field that supports every organ, every neuron, every cell. Picture a future where a clinician scans your field and sees the early signs of breakdown weeks before any symptom appears. Therapy restores your internal alignment instead of forcing biochemical reactions. Healing becomes the gentle act of stabilizing tension flows and rebalancing ripple dynamics.

No war. No violence. No side effects. Just the body returning to the coherence it was always meant to maintain.

This is not speculative mysticism. It is the logical next step once we understand the human body as a dynamic field system, not a biochemical machine. Medicine stops fighting nature and begins collaborating with the field that animates life itself.

Welcome to the future of healing—where coherence is the cure.

18.1 — The Body as a Coherent Ripple System

For centuries, biology has described the human body as a biochemical factory—a dense network of reactions, molecules, signals, and pathways. We mapped enzymes, sequenced DNA, traced neurotransmitters, and celebrated the triumph of chemistry as the language of life. Yet chemistry was never the source. It was the consequence.

At the most fundamental level, the body is not a machine of molecules. It is a coherent ripple system—a living architecture of stabilized patterns in the SVE field. Cells, tissues, and organs are not objects assembled like bricks; they are field configurations held in place by a delicate balance of tension, curvature, and alignment.

Your heart carries its own ripple signature. Your liver resonates with another. Your neurons pulse with intricate coherence chains that sustain memory, emotion, and perception. Even DNA—often treated as the master blueprint—is simply a physical imprint of a deeper field-level coherence logic.

When the field is stable, the body expresses health effortlessly. When coherence weakens, when local ripple structures distort, when tension accumulates or alignment breaks, the physical body begins to show symptoms. Illness, then, is not a biochemical accident. It is a misalignment—a region of the field that has lost harmony with the rest of the body's architecture.

Inflammation becomes a flare of ripple turbulence around a stressed region. Chronic disease emerges as long-term tension imbalance that the field can no longer compensate for. Mental health disorders arise from disrupted coherence across neural ripple networks, leading to chaotic signaling and emotional instability. Aging itself is the gradual decline in the body's ability to maintain coherence against constant ripple noise.

Chemistry reacts to these disruptions, but it does not cause them. Biochemical symptoms are the downstream expressions of deeper field instability—the body's attempt to compensate, adapt, or defend when its coherence is compromised.

Reframing the body as a ripple system shifts medicine from chasing symptoms to restoring alignment. Healing transforms from biochemical intervention into coherence stabilization.

The body is not a molecular battlefield. It is a symphony—and health depends on the stability of the field that conducts it.

18.2 — Ripple Diagnostics: Seeing the Field Before the Symptoms

Modern diagnostics waits for trouble. We detect disease only after it becomes visible—after cells mutate, tissues inflame, organs fail, or symptoms force us to act. Medicine today is reactive because it is blind to the deeper layer where illness begins: the ripple field that underlies the body's structure and function.

Ripple diagnostics changes that forever.

In a field-centric view of biology, the earliest sign of disease is not a biochemical marker or a physical abnormality. It is a drop in coherence—a subtle weakening of alignment in the local ripple architecture. Long before a tumor forms, before inflammation begins, before stress becomes chronic, the field shows disturbances: tiny turbulence, misaligned patterns, and micro-gradients that signal instability.

Field-mapping technologies—sensitive instruments capable of visualizing the body's SVE coherence—will detect what no scan or blood test can see today. They will reveal areas of tension accumulation, coherence fractures in neural networks, stress signatures in organ-level ripple patterns, early turbulence around aging or degenerating tissues, imbalances in emotional-field coupling, and hidden predispositions for disease long before onset.

Instead of seeing illness as something that suddenly appears, we begin to see it as something that develops—slowly, quietly, but measurably—in the field long before the body reacts. Ripple signatures become the new vital signs. A cardiologist will no longer rely solely on ECG waves but will examine the coherence profile of the cardiac field. A neurologist will not wait for cognitive decline but will map ripple stability across neural circuits. A mental health specialist will detect emotional strain as coherence distortion patterns far earlier than mood shifts or behavioral signs.

This reframing transforms medicine into a predictive science. We do not wait for damage; we anticipate it through field-scanning. We do not fight disease; we prevent misalignment before it expresses biochemically. We do not treat breakdowns; we maintain coherence proactively.

Ripple diagnostics gives humanity something it has never had before: the ability to see the future of health before the body pays the price.

18.3 — Coherence Therapy & Energy Stabilization

Once we understand that illness begins in the field and only later appears in the body, the logic of treatment undergoes a profound shift. Healing is no longer a battle against symptoms—it becomes the restoration of alignment, a gentle return of the body's ripple architecture to its natural state of harmony.

This is the essence of coherence therapy.

Instead of forcing biochemical reactions or suppressing physical responses, coherence therapy works directly with the SVE field. It tunes resonance, stabilizes tension flow, and smooths distortions before they cascade into disease.

Think of a musician adjusting the strings of an instrument. If one string falls out of tune, you don't attack it; you adjust it back into resonance with the rest of the instrument. Your body works the same way.

When an organ becomes misaligned, when neural pathways lose coherence, when emotional turbulence distorts the field, the goal is not to overpower the disruption but to re-harmonize the structure that sustains it.

This opens the door to therapies that are neither chemical nor invasive:

- **Field modulation** — gentle adjustments that restore smooth ripple flow across tissues and organs.
- **Resonance tuning** — aligning the body's local field patterns with their natural vibrational modes to reverse stress-induced distortions.
- **SVE equilibrium balancing** — reducing tension hotspots and preventing ripple cascades that lead to chronic disease.

These therapies operate on energy pathways, not molecular pathways. They do not force the body to change; they create the conditions in which the body can restore itself.

Where pharmaceuticals mask symptoms or intervene aggressively in biochemical cascades, coherence therapy addresses the origin—the field misalignment that causes the cascade.

Recovery, in this paradigm, is not a war won through stronger drugs but a return to coherence—the moment when the body's internal architecture once again flows with stability, clarity, and unified rhythm.

This is medicine stripped of violence and grounded in the true foundation of biology: the field that gives life its form, its function, and its potential for renewal.

18.4 — The Next Medical Paradigm: Integrated Field Health

The future of medicine will not be defined by smarter pills, sharper lasers, or more powerful surgical robots. The true revolution lies in a new operating system for health—one built not on chemicals and corrections, but on continuous coherence, personalized field insights, and early, elegant realignment.

In a field-centric medical paradigm, health is no longer defined as the absence of symptoms. It is the stability of one's ripple architecture—the long-term harmony of tension flows, coherence patterns, and alignment across organs, nerves, and emotional networks.

Continuous coherence monitoring becomes the cornerstone of next-generation healthcare. Just as we track heart rate or blood pressure today, tomorrow we will track ripple stability across major organs, coherence levels in neural pathways, tension hotspots forming before inflammation, emotional-field coupling patterns, and long-term ripple drift associated with aging or chronic disease.

Each person will carry a personalized field profile—a dynamic map of their body's coherence landscape that updates in real time. Healthcare shifts from occasional check-ups to continuous field supervision, quietly alerting us when misalignment begins long before symptoms appear.

This leads to a "correction-before-collapse" model—the medical equivalent of repairing the foundations of a building before cracks reach the walls. Chronic diseases become preventable. Mental health imbalances become detectable. Aging becomes manageable through regular coherence tuning. Regenerative medicine emerges naturally as the field learns to restore its own architecture.

The implications are profound. Chronic illness becomes a matter of field maintenance, not lifelong biochemical management. Mental health is treated through neural coherence repair, not emotional suppression. Aging slows as micro-instabilities are resolved before they cascade into cellular decline. Regeneration becomes feasible when damaged field structures are rebuilt, allowing tissues to follow suit.

In this paradigm, the body is no longer seen as a failing machine in need of constant repair. It becomes a living field system, capable of sustained renewal when its coherence is protected.

Medicine evolves from repairing the body to stabilizing the field that builds the body. This is the next medical revolution—quiet, elegant, preventative, and deeply aligned with the true nature of life.

Chapter 19 — Ripple Logic Machines

Beyond Quantum Computing

Coherence-based computation beyond qubits

For decades, quantum computing has been celebrated as the holy grail of future technology—the machine that would revolutionize cryptography, simulate entire worlds, and outperform classical computers by impossible margins. Yet the closer we get, the more quantum computing reveals a quiet truth that few want to acknowledge: it doesn't want to work.

Qubits collapse when touched. They decohere when cooled, heated, or even breathed near. They demand isolation chambers, vacuum enclosures, cryogenic temperatures, and endless error correction—just to remain stable for a few fleeting microseconds. This is not merely engineering difficulty. It is nature whispering a warning: *"You're trying to compute against the field."*

Quantum computing attempts to trap delicate superpositions inside a substrate that prefers coherence, not contradiction. It tries to build logic on top of a phenomenon the universe itself treats as temporary, unstable, and energetically expensive.

But what if we flipped the paradigm entirely? What if computing did not fight the universe's behavior, but used it directly? What if the substrate of the universe—the SVE field itself—could be the computer?

Imagine machines that compute by shifting field curvature, not by moving charges or maintaining fragile superpositions. Imagine processors where "bits" are patterns of coherence—stable, self-sustaining, and naturally aligned. Imagine memory stored as tension gradients, never degrading, never collapsing. Imagine a computer that grows more stable as its coherence increases.

No switching. No electricity. No cryogenics. No decoherence. Just ripple logic—computation written directly into the fabric of the field.

Quantum computing was the last dream of a particle-based worldview. Ripple logic machines are the first technology of a continuous universe. This chapter marks the turning point where information science stops struggling against instability and begins aligning with the architecture of reality itself.

19.1 — The Limits of Quantum Computing

Quantum computing was born from a beautiful idea: if a particle can exist in two states at once, then perhaps a computer built from such particles could process information in ways classical machines never could. It was an elegant dream—but one resting on a fundamental misunderstanding of how the universe actually maintains stability.

A qubit's supposed power lies in superposition—the ability to occupy multiple states simultaneously. Yet in a field-centric view of reality, superposition is not a natural "mode" of existence. It is a temporary coherence trick, a delicate balancing act between ripple patterns that the SVE field does not sustain willingly. The universe prefers stable curvature, not contradictory states stacked precariously atop one another.

This is why qubits collapse the moment they are touched. The instant you observe, disturb, or even approach the system, the field snaps back to a stable configuration. It is not "wavefunction collapse" caused by measurement—it is coherence collapse triggered by the field rejecting an unstable arrangement.

And the more qubits you add, the worse the problem becomes. Each additional qubit increases the demand for maintaining an unnatural coherence structure—a structure the field continuously tries to dissolve. This is why quantum systems refuse to scale beyond fragile prototypes.

To keep these unstable states alive, engineers deploy every imaginable remedy: cryogenic cooling, vacuum isolation, magnetic shielding, error correction codes, noise filtering, precision stabilization systems. Yet none of it works in principle, because none of it addresses the root cause: the qubit model is built on a state the universe does not want.

Decoherence is not a technical hurdle. It is the SVE field enforcing its natural behavior. Quantum computing attempts to compute by maintaining contradictions; the field resolves contradictions by collapsing them. No amount of engineering can overcome a substrate that fundamentally resists the architecture it is asked to support.

This is the hard ceiling of quantum technology. It is not an engineering challenge—it is a conceptual mismatch. Qubits operate against the field's natural dynamics. Ripple logic machines will operate with them.

19.2 — Computation as Field Reconfiguration, Not Particle Behavior

The moment we abandon the particle mindset, computing ceases to be a matter of moving charges, flipping transistors, or preserving fragile superpositions. In a continuous-field universe, information does not need a carrier—it only needs a pattern.

This is the foundation of ripple logic: a computing paradigm that treats information as field configurations, not particle states.

Instead of forcing electrons down wires or trapping qubits in delicate superpositions, ripple logic works by reconfiguring local curvature in the SVE field. A computation becomes the act of reshaping ripple geometry from one stable pattern to another.

Where classical computing relies on bits, and quantum computing relies on superpositions, ripple logic relies on coherence modes—naturally stable patterns of tension and alignment within the vacuum substrate. These modes are not fragile. They do not collapse. They are the preferred states of the field itself.

The SVE field supports them effortlessly because:

- Stable ripple patterns require no energy to maintain.
- Coherence strengthens itself through alignment.
- Curvature modes nest within larger tension structures.
- The field naturally resists noise and distortion.
- Small disturbances relax back into the nearest stable state.

In other words, ripple states act as self-correcting computational modes.

There is no need for superposition—a fundamentally unstable phenomenon—because ripple logic does not depend on ambiguity. It depends on stability. Just as the universe prefers coherent waves to chaotic noise, it prefers stable ripple modes to contradictory quantum states.

This makes ripple logic machines fundamentally robust:

- They operate at room temperature.
- They store information as tension patterns.
- They compute by shifting curvature.
- They maintain coherence naturally.
- They scale without collapsing into decoherence chaos.

Where quantum computing struggles to hold fragile states in a hostile environment, ripple logic builds its operations from the environment's own natural behavior.

This is the shift from computation as manipulation of particles to computation as reconfiguration of the field. In a universe made of ripples, information is not something we push—it is something we shape.

19.3 — Ripple Logic Gates, Gradient Memory & Coherence Pathways

In a field-centric universe, computation does not begin with switches, voltages, or binary states. It begins with the interactions of ripples—the same natural dynamics that govern waves, resonance, and stability throughout the SVE field. When these interactions are deliberately shaped, they become the building blocks of logic.

This is where ripple logic gates emerge.

A ripple-gate is not a transistor. It is a tension event—a controlled interaction between two or more ripple patterns that produces a predictable outcome. Where traditional gates rely on voltage thresholds and quantum gates rely on delicate entanglement, ripple-gates rely on constructive and destructive curvature flows.

- When ripple patterns align constructively, curvature amplifies → a logical "YES" condition.
- When they destructively interfere, curvature cancels → a logical "NO" condition.

The field itself performs the operation, and because the SVE substrate prefers stable solutions, every interaction naturally collapses into a coherent, noise-resistant result.

Information storage transforms just as radically. Instead of bits encoded as 0s and 1s or qubits encoded as fragile superpositions, ripple logic uses *gradient memory*—data stored as tension differences within the field. A gradient can persist indefinitely: the field maintains it without energy expenditure, acting like a perfectly stable memory cell that never leaks, never degrades, and never requires refreshing cycles.

Where silicon requires constant switching and quantum systems demand isolation, gradient memory simply exists as a natural equilibrium of the field.

Computation becomes massively parallel through *coherence pathways*: channels of aligned ripple flows that carry information not as particles or pulses, but as shifts in field geometry. Dozens, hundreds, or even thousands of operations can cascade simultaneously without interference or energy loss, because coherence pathways do not collide—they synchronize.

This makes ripple logic inherently efficient:

- No switching noise
- No electron motion
- No heat dissipation
- No cryogenic cooling

A ripple processor computes by allowing the field to find the next stable configuration—an act as natural as a drop of water settling into its calmest shape.

In this architecture, computation is effortless. Not because it is simple, but because it is finally aligned with the universe's own preferred way of organizing information.

19.4 — The Architecture of Ripple Logic Machines

Once computation is understood as the reconfiguration of the SVE field, the design of a computing device becomes both simpler and more profound. A ripple logic machine is not built from circuits, qubits, or switches—it is built from regions of controllable field curvature, carefully arranged so that ripples interact in predictable, stable ways.

A ripple processor works by shaping local field geometry. Each computational step is a shift of curvature, a reorientation of tension flows, a sculpted realignment of coherence patterns. It does not push electrons or trap quantum states. It simply nudges the field into a new equilibrium, allowing the solution to emerge naturally.

This architecture makes ripple machines inherently:

- **Scalable**, because coherence strengthens with size
- **Robust**, because the field self-corrects distortions
- **Room-temperature stable**, because no fragile superpositions exist
- **Energy-efficient**, because no charge transport is needed
- **Noise-resistant**, because coherence pathways reject turbulence

Where quantum computers demand extreme cold and classical computers waste energy as heat, ripple logic machines operate comfortably in everyday environments—quiet, cool, and consistent.

Their stability allows for architectures that were impossible under previous paradigms:

Ripple CPUs: General-purpose processors where curvature gates execute logic through tension interactions. They can handle classical tasks with extraordinary efficiency and perform complex, nonlinear operations natively without the overhead of binary switching.

Ripple Neural Processors: The first true analogues to biological intelligence: coherence networks that learn by reshaping their fields, forming long-term tension pathways—a computing model closer to the human brain than any silicon or quantum device could ever be.

Global Ripple Networks: Communication systems where nodes synchronize via field coupling rather than data transmission. A signal generated in one region of the SVE field can align distant nodes instantly through coherence, not speed. Such networks become the backbone of planetary-scale computation—a global field-computer that remains unified without wires, satellites, or optical channels.

Ripple logic machines overcome all limitations of quantum hardware because they do not fight the universe's substrate—they operate inside its natural modes. They are not fragile. They are not narrow. They are not unstable experiments.

They are the first computers that compute the way the universe itself computes.

19.5 — The Future of Information in a Field-Centric Universe

Computing as alignment, not manipulation

When computation shifts from particle manipulation to field alignment, the very meaning of information transforms.

In a particle-based worldview, information is something we push around: electrons through wires, photons through fibers, bits through processors. In quantum computing, it becomes even stranger—a fragile superposition that collapses if the universe breathes too loudly.

But in a field-centric universe, information is not an object and not a state waiting to collapse. Information becomes a property of coherence—of how ripple patterns align, stabilize, and evolve within the SVE substrate.

This is a profound shift. Instead of treating data as something external, stored on devices or encoded in silicon, we begin to see information as something the field itself knows through its own structure.

AI Reimagined Through Field Dynamics: Artificial intelligence ceases to be a brute-force model stacking layers and multiplying matrices. It becomes the art of shaping coherence networks that learn by reorganizing their field geometry—a process far closer to biological intelligence than anything silicon has ever produced. Ripple-based AI thinks not by calculation but by alignment, forming stable tension patterns that represent memory, prediction, intuition, and inference.

Simulation and Modeling Beyond Computation: In ripple logic machines, simulation is not approximation—it is resonance. To simulate a storm, a galaxy, a brain, or a protein, you shape the field into an equivalent ripple-state and let the dynamics unfold naturally.

Neuroscience Transformed: The brain is revealed as a coherence engine. Neurons are only the visible scaffolding of a deeper ripple intelligence. Understanding cognition becomes the study of how the neural field stabilizes patterns that produce thought, emotion, and awareness.

Planetary-Scale Computation: Once communication becomes coherence coupling, the entire planet can function as a unified computational organism—a network of ripple logic nodes synchronizing without transmission, satellites, or bandwidth limits.

A Continuous Computing Paradigm: Binary computing chopped reality into pieces. Quantum computing strained to hold contradictions. Ripple logic computing embraces continuity—the first system aligned with the universe's natural behavior.

When information becomes coherence, computing becomes alignment, intelligence becomes geometry, and technology becomes an extension of the universe's own logic. This is the future of information in a field-centric reality—fluid, unified, stable, and astonishingly powerful.

Chapter 20 — Energy Technologies

Aligned with Vacuum Structure

Equilibrium devices, ultra-efficient systems, ripple-energy extraction

Every energy technology humanity has ever built—from burning coal to splitting atoms—has been an act of violence. We crush, ignite, collide, tear apart, and force matter to surrender the energy bound within it. Our engines roar. Our reactors strain. Our grids hum under load. And every watt we extract comes with loss, heat, pollution, or catastrophic risk.

Why? Because for centuries we have tried to fight the universe, forcing it to release energy that did not want to be released.

But what if energy was never inside matter at all? What if the true reservoir has always been the structured vacuum field—the SVE substrate that gives rise to matter, motion, and coherence?

In a continuous universe, energy is not fuel. Energy is tension redistribution—the graceful movement of the field toward equilibrium. Every ripple, every gradient, every flow is already energy in motion.

If that is true, then the future of power does not demand reactors, turbines, or combustion. It demands alignment.

Imagine machines that run by cooperating with the vacuum, not resisting it. Imagine devices that operate silently and efficiently because they let the field do the work instead of forcing it. Imagine generators that last for decades because they are nothing more than stable equilibrium loops maintained within the SVE structure itself.

Ripple-energy extraction is not "free energy." It is field participation—drawing from the natural dynamics of a universe already alive with motion, coherence, and balance.

This chapter explores the moment when energy stops being extracted and starts being synchronized. A civilization that aligns with the vacuum no longer burns, breaks, or battles. It tunes, resonates, balances.

This is the beginning of energy without destruction—technology that moves with the universe, not against it.

20.1 — Energy as Tension Redistribution, Not Fuel Consumption

Humanity's understanding of energy has long been shaped by a simple but deeply flawed assumption: that energy lives inside matter, waiting to be released through burning, breaking, or collision. Coal, oil, gas, uranium—every major energy source we have relied upon comes from this belief that power requires destruction. We tear matter apart to free its "stored" energy, and in doing so, we inherit pollution, inefficiency, and enormous waste.

But this entire model collapses once the universe is seen correctly—not as particles locked inside particles, but as a continuous SVE field where energy is nothing more than tension flow within the substrate.

Energy is not a substance. It is a redistribution of curvature and ripple pressure in the vacuum field.

Every process we observe—heat, light, motion, electricity—is the field reorganizing itself to reduce imbalance. When we burn fuel, we are not "creating" energy; we are violently disturbing stable structures just to force the field to move in ways we can harness.

This is why conventional energy systems are inherently wasteful: they fight the field instead of aligning with it.

- Combustion forces chaotic ripple collapse.
- Turbines convert turbulence into partial order.
- Nuclear fission creates violent, short-lived gradients.
- Mechanical engines lose vast amounts to friction and heat.

In every case, we push against the natural behavior of the field—and the field pushes back.

The future of energy begins when we stop treating the universe as a vault to break open and start treating it as a dynamic equilibrium to collaborate with.

Enter equilibrium devices—technologies that operate by cooperating with natural tension flows instead of forcing new ones. These devices do not burn or collide matter. They work by:

- Stabilizing local gradients
- Harnessing natural ripple flows
- Tuning into existing field oscillations
- Guiding curvature relaxation
- Leveraging the vacuum's constant drive toward equilibrium

Instead of triggering massive disturbances to extract small pockets of usable power, equilibrium devices harvest the continuous motion that already exists in the field.

Energy becomes alignment, not violence. Power becomes participation, not extraction. And technology becomes gentle, efficient, silent—finally in harmony with the fabric of reality itself.

20.2 — Ultra-Efficient Technologies Through Alignment

When machines operate today, they waste most of their energy doing things the universe never asked them to do. Engines grind against friction, turbines churn through turbulence, motors wrestle with resistive forces, and processors bleed heat simply because electrons must be pushed through matter that resists their movement. Losses are not accidents—they are the cost of forcing the world to behave unnaturally.

But in a field-centric universe, efficiency is not achieved by better lubrication, stronger materials, or more advanced cooling. Efficiency is achieved by alignment—by allowing machines to operate with the natural coherence of the SVE field instead of against it.

When a device is tuned to the field's preferred mode of organization, something extraordinary happens:

- Friction drops dramatically
- Turbulence decreases to near zero
- Heat generation becomes negligible
- Energy flows stabilize instead of scattering

This is not magic. It is the consequence of *coherence coupling*—a state where the device's geometry and operation match the vacuum's own stable ripple patterns. In such a condition, the field does not resist movement; it supports it.

Imagine a motor whose rotor does not scrape against bearings but floats in a stabilized tension envelope—a frictionless, contact-free environment maintained by field alignment rather than magnets or air cushions. Imagine a turbine blade that slices through fluid with almost no drag because its ripple architecture minimizes turbulence instead of tearing through it. Imagine a processor that performs billions of operations without heating, because its computational transitions occur through curvature shifts instead of electron motion.

These machines do not "consume" energy in the traditional sense. They redirect tension flows already present in the field.

The transition from force-driven engineering to tension-flow engineering is the single greatest leap in efficiency in the history of technology. Motors, turbines, engines, processors, pumps, compressors—all can be redesigned to operate through alignment rather than brute force.

Where today's machines fight the vacuum, tomorrow's machines will glide through it—quiet, cool, and astonishingly efficient.

This is the era of technologies that do not strain against nature, but move in perfect coherence with it.

20.3 — Ripple-Energy Extraction: Harvesting Field Dynamics Safely

The greatest misunderstanding about vacuum-based energy comes from two extremes: those who dismiss it as impossible fantasy, and those who imagine a magical "free energy" machine that violates every conservation law ever written. Both positions fail because both assume the vacuum is empty.

In the UED view, the vacuum is not empty. It is a structured reservoir of organized potential—a dynamic field of ripples, gradients, and tension flows that constantly shift toward equilibrium. This reservoir is not infinite, but it is continuous and universal, providing a steady backdrop of energy movement that exists everywhere, all the time.

Ripple-energy extraction does not mean pulling unlimited power from nothing. It means partnering with natural field dynamics in a controlled, respectful way, much like windmills harness wind or solar panels harness sunlight—but on a deeper, subtler level.

The SVE field contains:

- Ripple-density variations created by matter, motion, and curvature
- Coherence wells formed by stable ripple structures
- Natural vacuum oscillations—tiny tension fluctuations that arise as the universe constantly relaxes toward equilibrium

None of this violates conservation. It is simply energy in motion, waiting for the right interface.

Ripple-energy extraction happens through:

1. **Resonance Coupling:** By tuning a device to the natural oscillation frequency of a local field region, it can absorb and convert tension shifts without external fuel. This is not tapping a "source"—it is syncing with ongoing field movement.
2. **Gradient Modulation:** Small curvature adjustments can create predictable tension flows, much like a water wheel placed in a gentle stream. The field itself does the work because it seeks smoother equilibrium.
3. **Local Equilibrium Shifts:** By altering ripple density in a controlled pocket, we create a temporary imbalance that relaxes by releasing useful energy—not explosively like chemical or nuclear reactions, but smoothly, continuously, like a stretched membrane returning to rest.

The key is respect for the field. Extraction must be stable, gentle, and harmonized with the vacuum's natural tendencies. When done properly, ripple-energy technology becomes the safest energy system possible—clean, silent, balanced, and fundamentally aligned with the structure of reality.

This is not "free energy." It is field-compatible energy, harvested without violence, without depletion, and without breaking the universe's rules.

20.4 — Equilibrium Generators & Field-Based Power Systems

The power systems of today rely on disruption—combustion that tears molecules apart, magnetic rotation that fights resistive drag, nuclear reactions that split or fuse atoms under extreme conditions. These machines work by forcing nature to release stored energy violently.

But in a continuous-field universe, the most powerful and stable systems are not those that break equilibrium—they are those that use equilibrium itself as the operating principle.

This is the foundation of **equilibrium generators** and **field-based power systems**: devices that operate by balancing, tuning, and stabilizing the SVE field instead of disturbing it.

Rather than pushing electrons or rotating heavy rotors, these generators rely on:

- **Stable field loops** — self-sustaining curvature patterns that circulate tension without friction or decay
- **Tension differentials** — engineered gradients that continuously relax into usable energy without combustion
- **Coherence engines** — systems that amplify alignment within the field and convert the resulting tension flow directly into electrical or mechanical power

These principles allow us to design power sources that operate at room temperature, with no exhaust, no radiation, no moving parts, and no need for constant fuel input. Once established, a stable loop can continue running indefinitely—not by violating conservation, but by maintaining a self-resetting balance inside the SVE substrate.

Imagine a generator built like a soft whirlpool in the vacuum field: a coherent curvature loop that continuously channels tension the way a river channels water. As long as the loop remains stable, energy flows naturally, silently, with almost no loss.

Imagine coherence engines that convert ripple alignment directly into usable power—a transformation as quiet as a heartbeat and as steady as a planetary orbit.

These devices do not "produce" energy in the traditional sense. They facilitate energy—allowing the field to undergo its natural relaxation cycles in a guided, harnessed form.

The result?

- Near-zero heat dissipation
- No wear-and-tear
- Unlimited operational lifespan
- Silent and vibration-free output
- Astonishing energy density without combustion or radiation

Equilibrium generators mark the shift from energy extraction to energy participation—the moment when power systems finally adopt the universe's own logic: stability, alignment, coherence, and effortless flow.

20.5 — A Field-Centric Energy Civilization

When energy is no longer something we rip out of matter but something we align with in the vacuum, civilization itself begins to change. The shift is not merely technological—it is civilizational, philosophical, and ecological. A society powered by combustion, extraction, and large-scale destruction cannot resemble a society powered by coherence, stability, and gentle field participation.

A field-centric energy civilization is built on abundance—not the reckless abundance of exploitation, but the natural abundance of a universe that maintains continuous tension flows everywhere. Energy stops being a scarce commodity and becomes a universal relationship between human technology and the SVE field.

Global Energy Implications: Power grids become obsolete. Each home, vehicle, industry, or settlement can carry its own equilibrium generator—a silent, small-scale system that draws from the vacuum's natural dynamics. No dependency on oil fields, coal mines, nuclear facilities, or massive solar farms. Energy decentralizes, stabilizes, and democratizes.

Industrial Transformation: Heavy industries that currently rely on extreme heat, pressure, and violent reactions will transition to field-aligned processes that operate at room temperature with near-zero waste. Instead of forcing materials into desired shapes, we will tune ripple architectures so matter reorganizes itself naturally.

Transportation Revolution: Vehicles powered by field gradients do not burn fuel, create emissions, or generate friction. They glide—quietly, efficiently—across land, air, space, and even water without turbulence or resistance. Noise pollution disappears. Carbon emissions become a historical footnote. Mobility becomes fluid, elegant, and drastically safer.

Climate Restoration: A civilization that no longer combusts matter automatically heals the climate. Pollution declines rapidly. Ecosystems regenerate. The planet stabilizes—not by sacrifice or austerity, but because we finally stopped treating the Earth as a furnace for energy production.

Planetary Infrastructure: Cities can be built in places previously considered inhospitable—deserts, mountains, open oceans, even floating altitudes—because energy and transportation no longer depend on resource geography. Humanity becomes free to distribute itself according to comfort, not constraint.

From Scarcity to Sufficiency: The old energy paradigm was defined by tension—political, economic, and environmental. The new one is defined by alignment. When energy comes from the natural coherence of the vacuum, civilization stops competing for power and begins collaborating with the field that sustains it.

In a field-centric world, energy is not something we take—it is something we tune into. And that shift creates not just new technologies, but a new kind of humanity.

Interlude — From Knowledge to Mindset

We have reached a quiet, powerful moment in this journey—the moment after possibility becomes clarity and before clarity becomes culture.

Part V was not merely a catalogue of futuristic technologies. It was a mirror held up to physics, revealing what becomes possible the instant we stop forcing the universe to behave like a machine of particles and start collaborating with the field that has always been underneath it.

We witnessed propulsion without propellant, computation without collapse, medicine without violence, materials without atoms, and energy without destruction.

But something deeper has changed. A shift has begun—not in laboratories, not in industries, but in the mindset itself.

For centuries, science advanced by slicing reality into smaller and smaller fragments: particles, forces, interactions, mechanisms, components. We believed understanding lived in dissection—that truth was hidden in the pieces.

Part V showed the opposite.

Truth is not hidden in pieces. Truth is written in the continuum.

The more we aligned with continuity, the more the universe opened itself. The more we tuned into field dynamics, the less the world resisted our questions. Every paradox dissolved. Every "mystery" revealed a simple pattern. Every impossibility became a design problem.

What changed was not physics—it was our mental lens. And that is the threshold we stand on now.

The next revolution will not be technological. The next revolution will be cognitive—a shift in how we teach, how we research, how we observe, how we interpret, and how we think. Because a field-centric universe demands a field-centric mind.

Students cannot continue memorizing particles that do not exist. Researchers cannot continue designing experiments based on outdated assumptions. Thinkers cannot continue drawing diagrams of a universe fragmented into mythical components.

The science of the future begins not with new machines but with new vocabulary, new intuition, new habits of thought.

Part VI is the bridge—from knowledge to mindset, from theory to culture, from understanding the field to thinking like the field.

Part VI — The New Science Mindset

"A new universe does not begin with new discoveries— it begins with new eyes."
"When the mind shifts from fragments to flow, science becomes effortless."

Short Overview of Part VI

Part VI closes the journey not with equations or technologies, but with the most transformative element of all—the mindset that makes a continuous universe thinkable.

After dissolving particles, forces, paradoxes, and technological limits, we arrive at a deeper realization: the future of physics will not be built by stacking new theories on top of old beliefs. It will be built by shaping a new scientific intuition, a new way of seeing the world, a new language for explanation and exploration.

This part guides that transition.

It begins with teaching. Physics education must evolve from memorizing particle charts to understanding ripple structures, coherence, tension flows, and field behavior. The next generation of students must learn the universe as a continuum, not a zoo of disconnected components.

It moves into research. Experiments change when we stop searching for objects and start studying interactions, gradients, alignments, and the dynamic behavior of the SVE field. What scientists look for—and how they measure it—must transform.

It culminates in a mindset. A continuous universe requires a mind that sees stability, not entities; patterns, not particles; relationships, not forces; fields, not fragments.

Part VI is not just a conclusion. It is an invitation to a new style of scientific thinking—fluid, unified, intuitive, and aligned with the true structure of reality.

This is the final evolution: from a fragmented worldview to a continuous consciousness of the cosmos.

Chapter 21 — Teaching Physics Without Particles

A new language for students and scholars

Walk into any physics classroom today and you will see a strange contradiction playing out. On the blackboard, the teacher draws electrons as tiny spheres, photons as arrows, quarks as colored dots, and forces as invisible threads pulling objects together. Students nod politely, memorizing the diagrams—but deep inside, something doesn't add up. Because none of it is real.

The universe is continuous, yet we teach it as if it were made of Lego blocks. We tell students that light is both a wave and a particle—as if contradictions are acceptable in science. We ask them to imagine an electron as a point object orbiting a nucleus, even though no such "orbit" exists. We preach forces as push–pull interactions, knowing full well that nothing in nature actually pushes or pulls across space. And then we wonder why physics feels difficult, mystical, or paradoxical.

The problem is not the students. The problem is the language. We inherited a vocabulary from an era that believed the universe was made of little balls. And even after a century of field equations, we still teach using metaphors that contradict the mathematics. How can students think clearly when the very words we use create illusions?

The future of physics education begins when we stop teaching particles that don't exist and start teaching the continuous field that actually does. This chapter is about changing the language—not to add new jargon, but to remove the old myths that have blocked understanding for generations.

If the universe is one field, our teaching must speak the language of unity, coherence, gradients, ripples, and alignment. Physics must be taught as a continuum, not as a catalogue of disconnected entities.

Teaching physics without particles is not a radical idea. It is simply teaching physics as it truly is.

21.1 — Retiring the Particle Vocabulary

Why particle-based teaching creates confusion

Physics education begins with good intentions—to make the invisible understandable. But somewhere along the way, simplicity turned into distortion.

We draw electrons as dots, photons as arrows, quarks as colored marbles, and forces as mysterious little lines "pulling" objects toward each other. These pictures are meant to help, yet they plant the very confusion students struggle with for the rest of their academic lives.

Because these images are not simplifications. They are misrepresentations.

The moment we teach an electron as a discrete object, we force students to imagine something that does not exist. The moment we describe a photon as a "packet," we encourage them to visualize a traveling bead of energy. The moment we talk about "force carriers," we imply a delivery system that behaves like particles in a mailroom.

This vocabulary creates an entire world of illusions—a particle zoo full of creatures doing mechanical actions. But when students progress and meet wave equations, field theories, and continuity-based mathematics, their intuition collapses.

They wonder:

- How can something be both a particle and a wave?

- How can "carriers" exist if fields act continuously?
- Why do forces magically travel across space?
- Why does the math show fields, but the textbook shows little balls?

Even advanced learners begin to accumulate internal contradictions. They memorize answers without understanding, repeat definitions without intuition, and treat physics as a puzzle of exceptions instead of a coherent description of reality.

The deeper issue is this: particle-language teaches fragmentation in a universe that is continuous. It forces the mind to picture broken pieces instead of unified flows. It trains students to imagine billiard balls in motion when the universe behaves like a single ocean of ripples.

The artistic diagrams must go. The billiard-ball analogies must go. The cartoon "photon arrows" must go. Not because they are childish, but because they teach the wrong universe.

Retiring particle vocabulary is not an upgrade. It is a necessary correction—the first step toward teaching physics in a way that aligns with how reality actually works.

21.2 — Introducing Field-First Intuition

Teaching through ripples, curvature, gradients, and coherence

If particle-language creates confusion, field-first intuition creates clarity—not because it is simpler, but because it mirrors the way the universe actually behaves. Students do not struggle in physics because the subject is inherently hard. They struggle because they are asked to imagine a reality that does not exist.

A field-centric approach flips this entirely.

Instead of showing particles "doing things," we show the field itself curving, rippling, aligning, stabilizing, and flowing through gradients. Suddenly, every phenomenon becomes intuitive.

Start with the SVE field as the foundation: Before teaching "objects," students learn that the universe is a continuous fabric—the Structured Vacuum Energy (SVE) field—still at equilibrium, dynamic where disturbed, coherent where stable, and expressive where gradients exist. This becomes the canvas on which all physics is drawn.

Teach ripples instead of particles: Where textbooks draw electrons as dots, we describe them as ripple-knots—regions where curvature becomes stable. Where photons are presented as little bullets, we show them as alignment waves, pure propagation without condensation. From the beginning, students learn stability and motion as patterns of the field, not the movement of solid objects.

Let gradients replace forces: Instead of memorizing "force laws," students see motion as a natural slide along tension gradients in the field. Gravity becomes the gentle fall through curvature. Electromagnetism becomes alignment behavior. Nuclear interactions become coherence conditions. Nothing pushes or pulls—everything flows.

Solve classical–quantum confusion at the root: When students understand that all phenomena arise from field behavior, they no longer ask why light is both a particle and a wave, why electrons "jump" orbits, or why measurement collapses states. Those paradoxes dissolve because the duality itself never existed.

Help students feel the field: A field-first curriculum trains intuition. Visualizing ripples, mapping gradients, understanding coherence as stability—these become natural mental habits. Students stop memorizing formulas and start sensing the dynamics.

Field-first intuition does not merely change understanding. It changes how the mind engages with reality—fluidly, continuously, coherently.

21.3 — A New Pedagogical Language

Replacing particles with patterns and interactions

A new physics cannot be taught using an old vocabulary. Language is not decoration—it is the mental architecture through which students construct meaning. If the words are wrong, the understanding will always be wrong.

The shift to a field-centric universe demands a shift to pattern-centric language—a vocabulary rooted not in objects, but in behaviors, coherences, and interactions within the SVE field.

Reframing the Fundamentals: The first step is to reinterpret every major term in a way that reflects continuity rather than fragmentation.

- An electron is no longer a point particle. It becomes a stable ripple-knot—a pocket of locked curvature held in coherence.
- A photon is not a packet flying through space. It becomes pure alignment propagation—the SVE field's tension moving as a traveling rhythm.
- A force is not a pushing or pulling entity. It becomes gradient behavior—the natural tendency of ripples to move along tension slopes.
- Energy is not a substance. It becomes tension redistribution—the universe relaxing, adjusting, and restoring equilibrium.

Each shift removes illusions and replaces them with genuine understanding.

A New Visual Grammar: To teach this vocabulary, we must redesign the diagrams students see.

No more balls orbiting nuclei. No more arrows representing photons. No more force vectors pointing like ghostly hands.

Instead, we introduce:

- Ripple maps showing how curvature stabilizes
- Tension landscapes where motion naturally flows
- Coherence diagrams that replace particle interactions
- Gradient fields that visually explain attraction and repulsion
- Alignment waves representing light propagation

These graphics are not artistic tricks. They are representations of how the universe actually behaves.

Ending Paradoxes at the Root: Once the vocabulary changes, paradoxes vanish.

Wave–particle duality evaporates because nothing was ever a particle. Quantum "spookiness" dissolves because coherence explains entanglement. Force-mediation myths disappear because fields require no carriers.

Students no longer wonder *How can both be true?* Instead, they see that continuity was the truth all along.

A new pedagogical language does not merely teach physics more clearly—it teaches it without contradictions, without metaphors that break, and without concepts that mislead.

It creates a generation who sees the universe as one flowing fabric—because that is exactly what it is.

21.4 — Training the Next Generation of Thinkers

Teaching scientific creativity through continuity

If physics is continuous, then the minds that study it must learn to think continuously. This is the real revolution—not in laboratories or equations, but in the way we cultivate intuition in the next generation.

For too long, students have been trained to memorize lists, charts, constants, formulas, and particle names—as if physics were a catalogue. But the universe is not a catalogue. It is a behavior. And behavior cannot be memorized. It must be understood.

From memorization to comprehension: Assessments must shift from rote questions—"State the postulates," "Define a photon," "List the force carriers"—to questions that examine intuition:

- What happens to ripple coherence near a strong gradient?
- How does stability emerge from locked curvature?
- Why does motion follow tension slopes?
- How would a ripple disturbance propagate through an SVE imbalance?

These are not trick questions. They are invitations to think like the universe—smoothly, continuously, logically.

Modeling phenomena as ripple dynamics: Instead of describing "collisions," "impacts," and "particle exchange," students learn to draw:

- Ripple interference patterns
- Tension maps around structures
- Gradient slopes that define natural motion
- Coherence zones that represent stability

When a student stops asking *"What do particles do?"* and starts asking *"What does the field want to do here?"* their entire worldview reorganizes. Physics becomes a living fabric, not a mechanical playground.

Teaching intuition early: tools that transform thinking Three practices anchor field-first intuition:

- **Coherence Mapping** — charting stable vs. unstable regions in the field to see why some configurations persist.
- **Gradient Visualization** — using contour-like drawings to reveal how motion naturally flows through tension landscapes.
- **Ripple Analogies** — translating complex interactions into

wave-like behaviors without relying on particle metaphors.

These tools build a mind that thinks in conditions, not components; in flow, not fragmentation.

Preparing future researchers: A unified universe requires unified thinkers—men and women who can look at any phenomenon and immediately perceive the underlying field behavior. They will innovate not by pushing the boundaries of old theories, but by seeing patterns others never noticed.

When the next generation learns continuity from the beginning, they will not "reform" physics. They will grow into the physics that was always waiting for them.

Chapter 22 — Research in the Field-First Era

How experiments change when ontology changes

Every scientific revolution begins the same way: not with a new instrument, not with a new discovery, but with a new assumption about what is real.

When we assumed the Earth was the center of the cosmos, we built instruments to track circular orbits that never existed. When we assumed matter was made of indivisible atoms, we designed experiments to smash things into smaller pieces. When we assumed forces acted across space like invisible hands, we searched for carriers, messengers, and mediators.

And for a hundred years, because we assumed particles were real, we built experiments that looked for particles. We fired beams, measured tracks, counted collisions, catalogued "events," and tried to reconstruct the mythical zoo we believed must be hiding inside matter.

But what happens when the underlying ontology changes? What happens when we realize that nothing in the universe is a discrete object—that everything is a ripple, a curvature, a tension flow within a continuous SVE field?

Suddenly, the particle detectors become ripple interpreters. The collision experiments become distortion studies. The "events" become artifacts of instability. The "discoveries" become temporary flickers misinterpreted as entities.

When the ontology shifts, the purpose of experiments shifts with it. The goal is no longer to find what is smallest, but to understand what is most fundamental:

- The field itself
- Its coherence and gradients
- Its stability thresholds
- The conditions that give rise to form

This chapter explores the moment when research stops looking for creatures in a zoo and starts mapping a living ocean.

Welcome to the field-first era—where experiments no longer chase particles, but observe patterns.

22.1 — The End of Object-Hunting Experiments

From detecting particles to observing behaviors

Modern experimental physics has been built on a simple belief: if something fundamental exists, we should be able to detect it. And for a century, "detection" has meant observing tracks in a chamber, hits on a sensor, spikes in a graph, or bright sprays of collision debris.

We treated these signatures as footprints of tiny objects—particles flying through detectors like microscopic bullets. Millions of experiments were designed around this assumption: smash, count, identify, classify. It was object-hunting at the smallest scale.

But this entire approach collapses the moment we accept that the universe is continuous.

Because in a field-first universe:

- A track is not a particle trajectory. It is the temporary path of a ripple disturbance stabilizing through an SVE imbalance.

- A hit is not the impact of an object. It is a spike in tension redistribution when a field disturbance interacts with a detector's structure.
- A signature is not the footprint of a "thing." It is the detector's interpretation of a changing coherence pattern in the field.

What we thought were objects were always behaviors.

The problem with particle-detection logic is not that it is inaccurate—it is that it teaches us to look for something that does not exist. When experiments are built to detect discrete events, the universe will always appear fragmented. We will always "discover" new particles whenever we reach new energies—not because particles exist, but because higher energies produce new kinds of temporary distortions that we mislabel as entities.

Reframing experiments through field behavior: Instead of asking: *"What particle did we just see?"* the field-first era asks: *"What behavior just occurred in the continuum?"*

This shift leads to a radically different experimental design:

- Instead of counting events, we map gradients across space and time.
- Instead of tracking hits, we measure tension flows within detectors.
- Instead of identifying particles, we observe coherence patterns and how they evolve after perturbations.

The experiment becomes a dance of the field, not a hunt for hidden creatures.

In the field-first era, physics stops collecting objects and starts understanding behavior—the true language of the universe.

22.2 — New Experimental Priorities

Measuring curvature, ripple stability, and SVE gradients

Once particles disappear from our ontology, the very targets of experimentation change. Instead of trying to capture mythical objects, the experimentalist begins to observe how the field behaves, stabilizes, distorts, aligns, and relaxes. The priorities of measurement shift from counting discrete "events" to studying the conditions and patterns that give rise to those events in the first place.

In a continuous universe, the key measurable quantities are no longer masses or collisions—they are ripple behaviors.

1. Ripple-density profiles: Every stable or unstable configuration in the SVE field has a unique distribution of ripple density. What particle physicists once interpreted as "mass" is better understood as the density and rigidity of a stable ripple-knot. Experiments begin mapping how these densities evolve when disturbed by energy inputs.

2. Stability lifetimes: Short-lived "particles" such as the W, Z, or Higgs were never particles—they were temporary distortions that quickly dissolved back into equilibrium. The meaningful measurement here is the lifetime of such distortions as a function of ripple tension and gradient stress. Stability time becomes more important than "identity."

3. Coherence thresholds: Instead of measuring interaction cross-sections, we probe the field's ability to maintain or lose coherence under stress. How much disturbance can a curvature pocket absorb before dissolving? How do these thresholds vary with energy and environment? These measurements reveal the true architecture of stable structures.

4. Alignment disturbances & gradient slopes: Motion, attraction, and repulsion become measurable as sliding down gradients. Instead of force diagrams, we map tension slopes in real time. Even gravitational behavior becomes visible as curvature expressions in the SVE field.

New Instruments for a New Ontology: A field-first era requires tools that can sense field behavior rather than "hits":

- **Coherence scanners** detect drops or surges in field stability.
- **Tension-meters** measure local SVE curvature shifts instead of electromagnetic field strengths.
- **Ripple-field analyzers** visualize ripple patterns, interference zones, and alignment propagation in real time.

These instruments do not observe objects—they observe the continuous choreography of the field.

Reinterpreting existing data: Even today's collider results begin to look different when viewed through UED logic: tracks become ripple relaxations, jets become gradient-healing cascades, decay signatures become coherence failures.

The data was never wrong—the ontology was.

By redefining what we aim to measure, we open the door to insights that were invisible under particle-thinking, yet obvious in a field-first universe.

22.3 — Redesigning High-Energy Physics

Why smashing matter becomes unnecessary (and misleading)

High-energy physics was built on a beautiful but flawed dream: if we smash matter hard enough, it will reveal its smallest pieces. This belief gave us the great colliders—machines of extraordinary precision and cost, capable of accelerating particles to near-light speeds and slamming them together in violent explosions.

For decades, we interpreted the resulting debris as evidence of "fundamental particles"—a zoo of exotic species born for a trillionth of a second before vanishing forever.

But in a continuous universe, this entire enterprise looks different.

Collisions don't reveal objects; they reveal distortions.

What a collider really produces is a momentary overstress in the SVE field—a region where the natural curvature is forced into chaos faster than the field can heal itself. The sprays, jets, tracks, and signatures are not fragments of matter. They are patterns of the field attempting to restore equilibrium after being violently displaced.

The so-called "particles" are simply short-lived ripple distortions, brief pockets of misalignment, flickers of unstable curvature that dissolve the moment the field regains balance. Treating these flickers as independent entities is like mistaking whitecaps for individual objects rather than expressions of an ocean under stress.

From fragmentation to controlled perturbation: The UED/RFD framework reframes high-energy research entirely. Instead of smashing matter to see what breaks, we begin perturbing the field gently and precisely to understand:

- Stability thresholds
- Coherence failures
- Ripple-density transitions
- Gradient collapses
- Curvature overload points

This is not destructive science—it is field diagnostics, more similar to medical imaging than to demolition.

Rethinking the W, Z, Higgs, and other "heavy modes": These were never objects. They were high-energy flickers—temporary distortions that appear only when the field is stressed beyond its ability to maintain smooth curvature. Their masses were not intrinsic. They were measurements of how much energy was needed to produce a particular type of instability.

By understanding them as flickers, not particles, we move toward a cleaner experimental paradigm—one that studies conditions, not creatures.

The new direction for high-energy physics: Colliders of the future won't try to break reality into pieces. They will aim to shape, tune, and probe the geometry of the field itself.

High-energy physics becomes not a search for the smallest things, but a deep exploration of how a continuous universe behaves under extreme tension.

22.4 — Precision, Coherence & Continuous Measurement

New methodologies for studying a continuous universe

The moment we stop chasing particles and start studying the field itself, experimental methodology transforms. Physics moves from counting discrete outcomes to tracking continuous evolution—the real behavior of a universe that never breaks, never jumps, never collapses, but flows.

The old approach relied on single numbers: a hit, a decay signature, a detection event, a probability spike. These were treated as primary data, when in truth they were the least informative fragments of a much richer phenomenon.

In a continuous universe, useful data is not a snapshot—it is the unfolding story across space and time.

Replacing discrete outcomes with continuous data streams: Instead of logging isolated events, experiments begin to record:

- How a ripple disturbance forms
- How it stabilizes or dissolves
- How gradients shift as energy spreads
- How coherence strengthens, weakens, or fractures

Sensors move from binary detection to continuous sensing—capturing the full choreography of the field. Just as seismology tracks waves through the Earth, field physics tracks ripples through the SVE continuum.

Real-time coherence monitoring: Coherence becomes the central measure of field health. Instruments monitor tiny shifts in stability with extreme sensitivity:

- Coherence rises → structure stabilizes
- Coherence drops → distortion or misalignment begins

Sudden changes in coherence explain what quantum physicists once labeled "random events." Nothing is random; coherence simply crosses a threshold.

Ripple-evolution tracking: Instead of asking *what appeared*, we study *how a ripple evolves*:

- Does it spread or contract?
- Does it merge with nearby disturbances?
- Does it oscillate, settle, or dissipate?

This replaces particle-decay diagrams with ripple-life diagrams—a far more accurate representation of reality.

Gradient-mapping across time and space: Motion becomes visible as gradients rise and fall like tides. A test mass sliding down curvature is no longer explained by "gravity acting on it," but by the evolving tension landscape around it. Mapping these gradients gives us a direct view of how the universe guides motion.

Ending quantum randomness and measurement paradoxes: Quantum physics declared that nature is probabilistic at the deepest level. But probability was only the language we invented because we lacked the ability to monitor continuous coherence.

Once we measure the field in real time, quantum uncertainty becomes:

- Coherence fluctuation
- Ripple sensitivity
- Threshold crossing
- Measurement-induced disturbance

The paradoxes dissolve because the discontinuities were never in nature—they were in our instruments.

The future of experimentation: In the field-first era, precision means continuity, and understanding means watching the field in motion, not freezing it into events.

This is the beginning of a science that finally observes the universe the way the universe behaves—smoothly, coherently, continuously.

22.5 — The Research Mindset Rewritten

Hypothesis-building in a world without fragmentation

Every revolution in physics ultimately becomes a revolution in thinking. The shift from particles to fields is not just a change in equations or experiments—it is a change in what scientists consider a good question.

For a century, research questions were framed as:

- What particle causes this phenomenon?
- What carrier mediates this force?
- What new object explains this anomaly?

This mindset rewarded fragmentation. Every unexplained signal became a candidate for a new particle. Every deviation from prediction invited an addition to the zoo. Every anomaly was treated as a missing entity instead of a misunderstood behavior.

But in a continuous universe, these questions dissolve.

The new question becomes: "What field condition produces this behavior?"

Suddenly, anomalies are not clues to hidden particles. They are windows into:

- Ripple misalignment
- Curvature overload
- Coherence collapse
- Local tension imbalance
- Measurement-induced disturbances

Where a particle physicist sees a "new resonance," a field-based researcher sees a temporary stability mode that flickers because the field cannot maintain curvature under stress. It is not an object; it is a reaction.

Rebuilding hypothesis logic: A field-first researcher no longer asks: *"What is this made of?"* but instead: *"What configuration of the field makes this appear?"* They stop thinking in terms of ingredients, and start thinking in terms of conditions. They are not hunting building blocks—they are mapping states of the field.

Interpreting anomalies as misalignments, not missing species: Consider a detector reading that deviates from prediction. The particle-era interpretation: *"Maybe there's a new particle we haven't discovered yet."* The field-era interpretation: *"Maybe the ripple coherence crossed a threshold we didn't account for."*

This mindset removes the reflex to add new mythical entities every time nature behaves unexpectedly. The anomaly becomes diagnostic, not discovery.

Training the scientific community for continuity: This shift requires a new culture: one that values pattern recognition over object identification, field intuition over particle metaphors, and coherence analytics over discrete-event counting.

Younger researchers must be trained to:

- Visualize curvature instead of collisions

- Think in gradients instead of forces
- Model evolution instead of decay
- Analyze stability instead of mass
- Seek relationships instead of entities

In this world, creativity is not the art of inventing new particles—it is the art of sensing new behaviors within a single continuous field.

The research mindset does not just evolve; it matures. It stops dissecting the universe and begins listening to it.

Chapter 23 — The Continuous Universe Mindset

Shifting from fragmentation to flow

There is a moment—quiet, subtle, almost unnoticeable—when a student of physics suddenly realizes that the universe is not made of pieces. No one announces it. No equation forces it. No experiment demands it. It simply arrives, like a shift in vision.

You look at a problem you've solved a hundred times—an electron in a potential well, light bending near a star, a photon moving through a fiber—and instead of seeing objects doing things, you see a field adjusting itself. A ripple stabilizing. A gradient relaxing. A coherence pattern shifting. A tension landscape guiding motion.

And in that moment, you understand something that cannot be taught explicitly: the universe has never been fragmented. We were.

The human mind evolved to survive in a world of objects—rocks, trees, predators, tools—so we instinctively cut reality into pieces to make sense of it. But the deeper we peer into nature, the less those pieces exist.

Physics has been fighting this mismatch for a century—hiding contradictions under the carpet of "duality," inventing particles to patch missing explanations, treating continuous phenomena as discrete events, and drawing diagrams that comfort the mind even when they betray the truth.

This final chapter is about unlearning that habit. It is about the transformation that happens when we stop forcing the universe into fragments and begin experiencing it as a continuous, flowing field. It is about the mental architecture required to live comfortably in a reality that never breaks, never jumps, never collides—but ripples, curves, aligns, stabilizes, and expresses itself as one seamless continuum.

The continuous universe is not just a theory. It is a way of seeing. A way of thinking. A way of being. And once your mind crosses that threshold, you cannot go back.

Welcome to the mindset that aligns with the universe itself.

23.1 — From Objects to Behaviors

Rewriting the deepest assumptions of scientific thinking

For centuries, physics has been driven by a powerful assumption—that the universe is built from things. Atoms, planets, particles, quarks, forces, bosons, even fields treated as objects. It is a worldview inherited from everyday survival: to navigate life, the mind learned to identify objects, separate them, categorize them, manipulate them.

This instinct once served us well. It helped our ancestors distinguish food from danger, tools from obstacles, allies from threats. But when we carried this object-based thinking into the foundations of physics, we unknowingly built our theories on a psychological bias rather than universal truth.

How object-thinking shaped the history of physics: Classical physics imagined billiard-ball particles colliding through forces. Quantum physics imagined point-like particles behaving mysteriously—sometimes waves, sometimes dots. Field theories still packaged disturbances as "force carriers" and "quanta." The Standard Model became a zoo because our minds demanded creatures to populate the microscopic world.

Every new anomaly invited another new object. Every unexplained effect became a missing piece. Every discovery announcement claimed a new particle had been "found," even though what was detected was always just a fleeting pattern of behavior. Object-thinking didn't just shape models—it limited them.

Why the mind clings to pieces: Fragmentation comforts us. It gives us a sense of control. A world with particles feels manageable, like a cosmic machine built from cogs and gears. Our intuition prefers separable entities because they are easier to imagine, draw, label, and memorize. But the universe isn't built to match our imagination. It is built to follow its own coherence.

From things to patterns: In a field-first worldview, nothing exists as a standalone "thing." Everything is an expression of the SVE field under certain conditions:

- Stability becomes an electron
- Alignment becomes a photon
- Curvature becomes gravity
- Coherence becomes structure
- Tension gradients become motion

We shift from asking, *"What is this made of?"* to asking, *"What behavior of the field appears here?"*

From entities to expressions: The electron is not a bead—it is a stable ripple-knot. The photon is not a particle—it is a traveling alignment. A force is not a push—it is a gradient flow.

When the mind stops searching for objects and starts perceiving behaviors, physics stops fighting itself and becomes effortless.

The universe has always been a field. It is time for our thinking to become one too.

23.2 — Seeing the World as Flow, Not Parts

Training perception to recognize continuity

To think in a continuous universe, the mind must first *see* continuity—not as an abstraction, not as a mathematical requirement, but as a natural way of experiencing reality.

We are surrounded by continuity every day, yet our training forces us to ignore it and fixate on objects instead of flows. A leaf floating on water is treated as the "object of attention," while the ripples guiding its motion go unnoticed. A bird gliding through air is admired for its form, not for the invisible gradient currents that carry it. A conversation is heard as words, not as the emotional ripples shaping tone and meaning.

The world is flow, but the mind keeps choosing parts.

Training perception to see gradients and ripples: In a field-centric worldview, we train ourselves—and our students—to recognize dynamics instead of objects:

- The subtle pressure differences that steer weather
- The contour of sound waves moving through air
- The shimmering patterns of heat rising from a surface
- The way water curves around stones
- The way crowds move not as individuals but as coherence-seeking patterns

These everyday examples mirror the SVE field: continuous, responsive, relational. Nothing jumps. Nothing teleports. Nothing behaves like a particle. Everything expresses tension seeking equilibrium. Once you start seeing these flows, the mind naturally shifts toward continuity.

Daily life as a teacher of field behavior: Consider:

- A river: not objects moving in water, but water restructuring

itself around objects
- Wind: not air particles traveling, but pressure gradients rebalancing
- Music: not discrete notes, but coherent vibrations shaping emotional response
- Human emotion: not isolated feelings, but field-like waves rising and settling, influencing others through resonance

The physical, the biological, the emotional—all reveal the same principle: flow, not fragmentation.

Building comfort with unbroken reality: A universe that never breaks can feel strange at first because our education taught us to expect boundaries, pieces, collisions, edges. But when students and researchers practice visualizing everything as gradients, ripples, coherences, and relaxations, the continuous universe becomes intuitive.

They stop waiting for "events." They start watching evolution. They stop picturing particles. They start sensing patterns. They stop imagining forces. They start recognizing tension realignment.

With this shift, the universe stops being a collection of parts and becomes a living flow. A fabric. A field.

Exactly as it is.

23.3 — The Cognitive Tools of Continuity

Curvature thinking, gradient reasoning, coherence intuition

To think in a continuous universe, you need more than new concepts—you need new cognitive tools. Old physics trained us to think in straight lines: *A causes B, this pushes that, one object collides with another.* A world of billiard balls and impulses.

But a continuous field does not operate through cause–effect chains. It operates through conditions—curvature, tension, alignment, coherence, imbalance. Events are not triggered; they unfold.

Curvature Thinking — seeing reality as shaped space, not moving pieces: Curvature thinking replaces the habit of imagining particles in motion.

Instead of asking: *What force makes this object move?* the continuous mind asks: *What curvature guides its natural path?*

In gravity, curvature thinking is obvious—objects follow the shape of space. But in UED, all interactions express curvature: electrons stabilize where curvature locks, photons propagate where curvature aligns, ripples travel where curvature slopes allow flow. The question is never who pushed what, but how the field configured itself.

Gradient Reasoning — motion as response, not command: Once you stop imagining forces, motion becomes intuitive. Everything moves according to gradients—down slopes of tension, into regions of lower distortion, toward equilibrium.

Gradient reasoning teaches that nothing is compelled; everything simply follows the easiest path the field offers. This dissolves all "action-at-a-distance" myths because behavior is local, continuous, automatic.

Coherence Intuition — stability as a field property: Coherence intuition is the ability to sense whether a configuration of the field is stable or unstable.

You stop asking: *Why did this particle decay?* and instead ask: *Which coherence threshold was crossed?*

This single shift eliminates quantum randomness, because decay is no longer probabilistic—it is a response to rising tension or falling stability.

Replacing cause–effect with condition–consequence logic: The universe does not operate through collisions. It operates through conditions expressing consequences:

- Alignment → propagation
- Curvature → motion

- Gradient → behavior
- Coherence loss → transformation
- Tension accumulation → ripple formation

Events stop being "interactions between objects" and become field expressions responding to imbalance.

Thinking in alignment, balance, relaxation, flow: The continuous mind learns to ask:

- What is aligning?
- What is relaxing?
- What is rebalancing?
- What is flowing?
- What is settling into coherence?

This way of thinking is not merely conceptual. It becomes a habit—a new scientific instinct.

When these cognitive tools take root, the universe stops looking mysterious. It starts looking inevitable.

23.4 — Living With a Unified Worldview

How continuity changes creativity, problem-solving, and perception

A continuous universe is not just a scientific idea—it is a way of experiencing reality. Once you stop seeing the world as a collection of parts and start perceiving it as one flowing field, something remarkable happens: your creativity changes, your problem-solving changes, your understanding of people changes, your inner experience changes.

You begin to live in coherence.

Scientific Implications — theories without fragmentation: A unified worldview makes scientific creativity far more powerful and far less chaotic.

Instead of building theories by stitching fragments—gravity theory here, quantum mechanics there, particles added to patch anomalies, constants inserted to fix mismatches—a continuous mindset searches for one principle expressed in many conditions.

Researchers no longer add layers of explanation. They remove them. They simplify. They search for the underlying pattern instead of inventing new entities to explain old contradictions.

This creates theories that are elegant, predictive, and free of paradox.

Technological Implications — engineering through shaping, not forcing: Field engineering becomes natural when you stop picturing machines as devices that push, pull, burn, or collide.

You start thinking like the universe:

- Shape gradients, don't fight them
- Tune coherence, don't break it
- Align with the field, don't overpower it

Propulsion becomes surfing. Computation becomes reconfiguration. Energy becomes tension balancing.

The mindset shifts from control to collaboration with the underlying structure of reality. Technology becomes gentler, more efficient, more elegant—because it mirrors the universe's own design language.

Personal Implications — coherence instead of conflict: A unified worldview changes how you understand people too. Emotions become ripples, not attacks. Relationships become gradients, not battles. Disagreements become coherence mismatches, not enemies. Inner turmoil becomes tension that needs alignment, not punishment.

You begin to interpret human behavior the way you interpret physics: continuously, compassionately, coherently. You stop seeing broken pieces in people and start seeing unstable ripples seeking equilibrium.

Seeing the world as one field: When the mind finally grasps that reality is not made of parts, but of one coherent field expressing itself in different conditions, you begin to live differently.

You perceive connection everywhere. You find simplicity beneath complexity. You see patterns where others see problems. You sense flow where others sense conflict.

The unified worldview becomes a way of thinking—and then a way of being.

And once this mindset settles, you no longer struggle to understand the universe. You begin to move with it.

23.5 — The Continuous Mind as the Future of Science

Why the next breakthroughs require a new way of thinking

The greatest scientific breakthroughs of the next century will not come from bigger machines, deeper tunnels, or more expensive detectors. They will come from a new kind of mind—a mind that no longer fragments reality into imaginary pieces, but understands the universe as a continuous, coherent, living field.

UED and RFD are not merely new theories. They are invitations—an invitation to evolve the very architecture of thinking itself.

Why new models are not enough: History shows that when old mindsets survive, new theories get twisted to fit outdated intuition. Quantum mechanics was radical, yet we forced it into particle metaphors. Relativity was elegant, yet we surrounded it with gravitational "forces." Field theories unified interactions, yet textbooks still taught force carriers and particle exchange.

UED/RFD is too unified, too elegant, too complete to tolerate old cognitive habits. To understand a continuous universe, we need a continuous mind—one trained to think in ripples, coherences, curvatures, gradients, and flow.

How a field-centric mindset dissolves paradoxes automatically: Most scientific paradoxes do not arise from nature. They arise from the observer's mental model.

- Wave–particle duality vanishes when you stop believing in particles.
- Quantum randomness disappears when coherence thresholds are understood as field instabilities, not probability magic.
- Entanglement ceases to be "spooky" once coherence is recognized as a single structure, not two linked objects.
- Gravity needs no force when curvature naturally guides motion.

The continuous mind does not "solve" paradoxes. It never creates them.

Preparing humanity for the next stage of clarity: The next leaps in science will require:

- Researchers who think in conditions, not components
- Engineers who shape fields, not force matter
- Educators who teach flow, not fragmentation
- Students who grow up visualizing coherence, not collisions

Such minds will see connections where earlier generations saw confusion. They will design technologies that align with the universe instead of battling it. They will build theories that feel natural, not forced.

A mindset aligned with the universe: The continuous mind is not a luxury. It is the future operating system of science—a way of thinking that matches the structure of reality itself.

When humanity adopts this mindset, physics will not just advance. It will unfold—smoothly, coherently, the way the universe always has.

Interlude — When Understanding Becomes Identity

We have arrived at a quiet turning point—not in theory, not in technology, but in ourselves.

Part VI was more than the conclusion of a book. It was the moment where the universe stopped being a subject we study and became a lens we think through.

Something shifts inside you when you stop seeing electrons, photons, quarks, forces—and start seeing ripples, tensions, gradients, coherences. Something changes when you stop imagining collisions and events and begin perceiving continuous evolution. Something transforms when the universe stops being a set of problems and becomes a single, flowing expression reshaping itself moment by moment.

This is not merely understanding. It is identity. A new mental habit. A new way the mind exists within reality.

After living with this continuous worldview long enough, you notice subtle but profound differences:

- You think in flows instead of fragments.
- You sense patterns instead of objects.
- You explain phenomena with alignment and tension, not with collisions or carriers.
- You interpret anomalies as transitions, not as missing pieces.

- You feel the universe as something alive, not assembled.

Part VI is the moment when physics stops being a subject and becomes a mode of awareness. It is the point where clarity stabilizes, where contradictions vanish, where intuition and reality finally agree.

And now, with the mindset reset, something deeper becomes possible—something this book was quietly preparing you for from the first page:

A way to close the Particle Age and welcome the Field Age with complete confidence.

Let us now move to the final closing section—the one moment in which the entire journey comes together as a single, coherent truth.

Closing Section

"The universe is a single continuous field. There are no pieces. There are only ripples."

The Universe Was Never Divided

There comes a moment in every intellectual journey when you realize that the truth was never complicated—only our ways of seeing were.

This book has walked through the deepest corridors of physics, not to tear down the past, but to reveal a simple, continuous foundation that was always present beneath the noise: the universe has never been made of parts.

Particles were stories we told ourselves to navigate complexity. Forces were metaphors we invented to avoid confronting continuity. Collisions, interactions, exchanges—all clever attempts to describe a world we assumed must be fragmented.

But reality is more elegant than our imagination.

The universe was never divided. We were.

From the smallest ripple of an electron to the birth of a star, from quantum flickers to cosmic curvature, from consciousness to the fabric of motion—every phenomenon you have ever known is simply the SVE field seeking equilibrium, coherence, and flow.

Nothing appears suddenly. Nothing travels as a hard object. Nothing "jumps" between states. Nothing pulls or pushes across space.

It is all one movement—one dance—one unbroken continuum expressing itself through gradients, ripples, alignment, stability, collapse, relaxation, and renewal.

We mistook the rhythms for the dancers. We mistook the flickers for particles. We mistook the gradients for forces. We mistook the distortions for objects. We mistook the behaviors for things.

And when we let go of that illusion, the contradictions evaporated. Duality dissolved. Paradoxes disappeared. Quantum "weirdness" became natural. Gravity became intuitive. Cosmology became transparent. Technology became visionary. Science regained coherence.

Reclaiming simplicity is the true revolution.

The end of the Particle Age is not the end of physics. It is the end of a particular vocabulary, a particular worldview, a particular psychological habit.

The real physics—the field-centric physics—is just beginning.

We now understand that:

- An electron is a stable curvature-knot.
- A photon is pure alignment propagation.
- A neutrino is near-zero distortion on a coherent glide.
- Heavy modes are stressed flickers, not entities.
- Gravity is slope, not pull.
- Forces are illusions created by gradients.
- Energy is tension seeking balance.
- Motion is the field relaxing itself.
- Time is consequence unfolding.

And the universe—this grand, dynamic, breathing continuum—moves without fragmentation, acts without collisions, creates without pieces.

The next leap is not technological. It is cognitive.

A continuous universe demands a continuous mind.

A scientific culture that thinks in:

- Coherence instead of conflict

- Gradients instead of forces
- Ripple evolution instead of particle interactions
- Stability thresholds instead of "decays"
- Conditions instead of causes
- Flow instead of fragmentation

This is not merely a new chapter in physics. It is a new chapter in human clarity.

The Universe Was Never Divided. It is time for our thinking to catch up.

As you turn the final page, carry this truth forward—into your research, your teaching, your problem-solving, your creativity, your perception of nature, your understanding of others, and your own inner world.

The universe is one field, expressing itself in infinite forms. And the moment you begin seeing it that way, your mind aligns with the very fabric of reality.

This is the real revolution: a shift from knowing about the universe to thinking like it. A shift from fragmentation to flow.

Epilogue — A Letter to the Next Generation of Scientists

To the young scientist who will read this in a classroom, a laboratory, a hostel room late at night, or perhaps years from now when the world of physics has already begun to shift—

This letter is for you.

You are stepping into science at a time of tremendous transition. The foundations are trembling—not because they are weak, but because we finally have the courage to remove what never belonged.

For a century, we trained you to think in particles, to imagine the universe as a museum of objects, each with its own personality, each performing scripted actions. We taught you dualities instead of clarity, forces instead of gradients, collisions instead of conditions. We asked you to memorize diagrams we knew did not represent reality.

And yet—despite all the myths, all the metaphors, all the comforting simplifications—you still asked deeper questions. You wondered why nothing made intuitive sense. You sensed the inconsistencies early. You noticed that equations spoke a language the textbooks avoided. You felt the fracture between what you were told and what the universe whispered.

This book is my way of telling you: your intuition was right.

The universe was never divided. The fragmentation was human. The continuity is real.

As you carry science forward, here is what I want you to remember—

1. **Never mistake the map for the terrain.** Diagrams are tools. Models are conveniences. But reality is always richer, smoother, more coherent than our symbols.
2. **Trust the unity of nature.** When a theory feels patchy, fragmented, forced—step back. The universe does not operate in fragments. If your thinking becomes continuous, contradictions dissolve on their own.
3. **Develop intuition, not just knowledge.** Memorizing is not understanding. Understanding is not feeling. Feel the field. Feel the gradients. Feel the flow of reality. When you do, equations become reminders, not answers.
4. **Question even the sacred.** Not with arrogance, but with clarity. Every great leap in physics came from someone who refused to accept that yesterday's vocabulary should define tomorrow's truth.
5. **Let curiosity be your rebellion.** The world will try to confine your thinking through tradition, authority, or precedent. Curiosity is the force that frees the mind from inherited cages.
6. **Serve the truth, not the trend.** Science is not a competition of prestige. It is a commitment to clarity—even if clarity demands letting go of frameworks that built entire careers.
7. **Remember that humility is power.** The universe is vast, and you are privileged to witness even a fraction of its mathematics. Let awe guide you. Let wonder steady you.
8. **Think beyond fragmentation.** Your generation is the first that can begin its scientific journey without inheriting the particle illusion. Where we fought contradictions, you will feel coherence. Where we debated interpretations, you will see flow. Where we built bridges between competing theories,

you will build a unified landscape from the beginning.

9. **Build the tools we never had.** Coherence scanners, ripple analyzers, gradient-mapping systems, field-alignment technologies—these will not arise from the old worldview. They will arise from you.
10. **And above all—stay human.** Science does not progress through equations alone. It progresses through minds that are open, courageous, and willing to see reality without distortion.

As you step into your journey, know this: you are not inheriting a broken physics. You are inheriting a physics that is ready to evolve—a physics that finally mirrors the elegance of the universe itself.

The next age of science will not be built by those who defend the past, but by those who listen to the universe with fresh eyes and undivided minds.

Walk forward with clarity. Walk forward with humility. Walk forward with courage.

The universe was never divided. Now, finally—neither are you.

— Sandeep Chavan,

Independent Researcher, Founder of UED/RFD, India

Appendices

"The Field Age is not rebellion against physics — it is rebellion against misinterpretation."

Appendix A — Ripple Field Glossary

A *clear, student-friendly, researcher-ready vocabulary for UED/RFD* This glossary is designed for all levels—beginners, advanced students, researchers, and professionals—who want a clean, intuitive language for a continuous-field universe. No particles, no forces, no contradictions—only clarity.

Core Foundations

1. **Structured Vacuum Energy (SVE)** — The foundational field of the universe. Still when undisturbed, dynamic when rippled. Matter, motion, and light arise from changes in its curvature, tension, and coherence.
2. **Ripple** — A disturbance in the SVE field. Not a wave on a medium, but the medium itself expressing imbalance. All phenomena are ripple behaviors.
3. **Ripple-Knot (Electron Analogue)** — A stable, self-sustaining curvature pocket in the SVE field. What mainstream physics calls an "electron" is actually this stable ripple knot.

Field Dynamics

1. **Curvature** — How the SVE field bends or warps due to tension. Curvature guides motion naturally—no forces

required.

2. **Tension** — The "stress" stored in the field when displaced from equilibrium. This is the real meaning of energy.
3. **Gradient** — A slope in the field's tension or curvature. Objects move along these gradients automatically—what we once called forces.
4. **Coherence** — A measure of stability and order in ripple structures. High coherence = stability; low coherence = instability, decay, transformation.
5. **Alignment** — When the field's ripples move in sync. Light (photon) is simply the field propagating alignment.

Measurement & Stability

1. **Ripple Density** — How tightly packed ripple patterns are in a region. Higher density → heavier/stable modes; lower density → lighter/freely propagating modes.
2. **Stability Threshold** — The minimum coherence needed for a ripple structure to survive. Below this, it dissolves back into equilibrium.
3. **Distortion Flicker (Heavy Mode Analogue: W, Z, Higgs)** — A short-lived curvature disturbance created during high-energy stress. Not a particle—just a temporary imbalance that collapses quickly.
4. **Field Relaxation** — The natural process where the SVE returns to equilibrium. All motion, decay, and evolution arise from this relaxation.
5. **Gradient Flow (Gravity Analogue)** — Movement along curvature slopes. Gravity is simply tension redistribution in the field—not a force, not an attraction.
6. **Coherence Collapse (Quantum "Decay")** — When a ripple-knot loses stability and dissolves. No randomness—just coherence crossing a threshold.

7. **Alignment Wave (Photon Analogue)** — A ripple of pure direction and tension change in the field. Travels without mass, without condensation—the cleanest expression of propagation.

Experimental Vocabulary

1. **Ripple Signature** — The visible outcome of a ripple interacting with a detector. Tracks, jets, hits—signatures, not particles.
2. **Gradient Map** — A visualization of tension slopes in the field. Shows how motion and behavior naturally flow.
3. **Coherence Map** — A representation of stable vs. unstable regions. Used for studying ripple stability, diagnostics, and field engineering.
4. **Ripple Interference** — When two disturbances overlap, creating local boosts or cancellations. Equivalent to wave interference, but without a separate medium.
5. **Field Condition Logic** — Understanding events not through cause–effect, but through what the field is conditioned to do under given tensions.

Behavior & Stress

1. **Equilibrium Point** — A configuration where tensions balance. The universe constantly moves toward these points.
2. **Misalignment Stress** — Strain that builds when ripples are out of sync, often leading to flickers, turbulence, or instability.
3. **Ripple Turbulence** — Chaotic motion in the field when disturbances pile up. Misinterpreted in experiments as "random events."
4. **SVE Overflow** — When the field is overstressed (high-energy collisions), creating temporary distortions that appear as

heavy modes.

5. **Continuity Law** — The principle that the universe never breaks or jumps—it evolves continuously in ripple and curvature.

Applied Concepts

1. **Ripple Logic (Computing)** — Processing information by shifting ripple states, not flipping bits or maintaining quantum superpositions.
2. **Tension Redistribution (Energy Transfer)** — Energy is not a substance—it is the rebalancing of field tension from one region to another.
3. **Curvature Corridor** — A shaped gradient pathway allowing movement with minimal energy. Basis of future propulsion and gravity engineering.
4. **Coherence Engine (Power Source)** — A device that maintains a stable ripple loop, producing usable energy through continuous tension flow.
5. **Field Architecture** — Designing materials, machines, and technology by shaping the SVE field rather than manipulating matter.

Translation & Engineering

1. **Ripple Signature Catalogue** — Mapping of all known "particle signatures" to the actual field behaviors they represent.
2. **Field Interpretation Framework** — Rule-set for translating experimental data from particle-language to ripple-language.
3. **Equilibrium-Based Propulsion** — Movement achieved by shaping gradients instead of expelling mass.
4. **Alignment Resonance** — When ripple patterns amplify each

other, leading to high-efficiency transmission or stabilization.

5. **Dissolution Path** — The predictable way unstable structures return to equilibrium.
6. **Field Surfing** — Traveling along controlled SVE gradients—the future of non-propellant navigation.
7. **Ripple-Boundary Effect** — Where changes in coherence create temporary "edges" in the field—misinterpreted as particle boundaries.
8. **Stability Peak** — The highest state of coherence a ripple structure can reach before transitioning into a new mode.
9. **Gradient Asymmetry** — When one side of a field is more relaxed than the other, creating directed motion.
10. **Field Singularity (No Infinite Density)** — A point of extreme curvature—not infinite density—where coherence fully compresses before rebalancing.

Appendix B — UED/ RFD Principles

A master blueprint of Universal Energy Dynamics (UED) and Ripple Field Dynamics (RFD)

Universal Energy Dynamics (UED) — The Core Principles

UED explains the universe not through particles and forces, but through one continuous field—the **Structured Vacuum Energy (SVE)**—and the dynamics that emerge when this field is disturbed.

1. The Universe Is One Continuous Field Everything—matter, light, gravity, motion, time—arises from the behavior of the SVE field. Still when undisturbed, dynamic when rippled.

2. Disturbances = Ripples All phenomena are ripple behaviors: curvature, alignment, oscillation, collapse, coherence, gradients. Nothing exists as a discrete object.

3. Stability Creates "Things" Electrons, protons, atoms, and structures are not objects but stable ripple-knots where curvature locks into long-lived patterns.

4. Heavy Particles Are Flickers W, Z, Higgs, and other short-lived "particles" are high-stress distortions—temporary flickers as the field restores equilibrium.

5. Light = Alignment Propagation A photon is not a particle. It is a traveling alignment of field tension—pure propagation without condensation.

6. Gravity = Gradient Flow Gravity is not a force or spacetime curvature. It is the SVE field aligning with tension gradients. Motion naturally follows these slopes.

7. Energy = Tension Redistribution Energy is not a substance. It is the redistribution of tension as the field moves toward equilibrium.

8. Motion Is Field Relaxation Objects move because the field rebalances. No pushing, pulling, or force exchange—only continuous adjustment.

9. Time = Change, Not a Dimension Time measures how ripple conditions evolve as tension redistributes. It is not an independent dimension.

10. Equilibrium Governs Everything From quantum transitions to cosmic structures, the field always moves toward balance. This is the **Consequence Path**.

Ripple Field Dynamics (RFD) — The Behavioral Rules

RFD describes how the field behaves when disturbed—the fine-grained mechanics of ripples.

1. Ripple Formation — Local tension crosses a threshold and equilibrium breaks.

2. Ripple Propagation — Ripples move until smoother curvature is found—no carriers, no particles.

3. Ripple Interference — Ripples combine or cancel depending on alignment, phase, and tension.

4. Coherence — A ripple survives if internal alignment exceeds environmental turbulence.

5. Charge as Asymmetry — Positive/negative charge is ripple curvature asymmetry influencing gradient flow.

6. Mass as Rigidity — Mass is ripple density and rigidity—the resistance to curvature change.

7. Quantum Transitions = Coherence Collapse — Systems "jump" only when coherence thresholds are crossed.

8. Entanglement = Shared Coherence Region — Two ripples behave as one because they share a coherence zone.

9. Black Holes = Density Collapse Zones — Regions of extreme ripple density absorbing disturbance without emitting alignment.

10. Cosmology Without Exotic Entities —

- Dark matter = misread ripple density gradients
- Dark energy = large-scale tension relaxation
- Inflation = early turbulence smoothing out

The 12 Laws of UED/RFD

1. The universe is a continuous field.
2. All events are ripple behaviors.
3. Stability creates structure.
4. Flickers are not entities.
5. Light is alignment propagation.
6. Gravity is gradient flow.
7. Energy is tension movement.
8. Motion is equilibrium-seeking.
9. Time is change.
10. Coherence defines stability.
11. Misalignment defines transformation.
12. Equilibrium governs all consequence.

Appendix C — Field Reinterpretation

Of All Particle Signatures

A *one-to-one reinterpretation of particle discoveries using UED/RFD logic*

Why This Appendix Matters

For over a century, experiments have been interpreted as if detectors captured objects flying through them. But detectors never saw particles. They saw **patterns of field disturbance**—the consequences of SVE imbalance, ripple interactions, and curvature dynamics.

This appendix provides a direct reinterpretation of all major particle "observations" into their true field explanations.

C.1 — Electrons (Stable Ripple-Knots)

- **Mainstream View:** Tiny negatively charged particles orbiting nuclei.
- **Observed Signatures:** Spots on detectors, trails in cloud chambers, absorption lines.
- **Field Interpretation:** Electrons are stable ripple-knots—pockets of sustained curvature with high coherence, stable ripple density, and curvature asymmetry (interpreted as charge). Their signatures arise from field relaxation, not

particle impacts.

C.2 — Photons (Alignment Waves)

- **Mainstream View:** Massless particles carrying energy.
- **Observed Signatures:** Photoelectric effect, interference fringes, Compton scattering.
- **Field Interpretation:** Photons are traveling alignments of SVE tension—ripples without condensation. Signatures reflect coherence transfer, not particle impact.

C.3 — Neutrinos (Minimal-Distortion Ripples)

- **Mainstream View:** Ghost-like particles with near-zero mass.
- **Observed Signatures:** Missing energy in beta decay, Cherenkov light.
- **Field Interpretation:** Neutrinos are near-zero curvature distortions—soft ripple slips restoring equilibrium. Detectors catch them only when low-density ripples encounter resistance.

C.4 — Quarks (Internal Distortion Modes)

- **Mainstream View:** Fractionally charged particles inside nucleons.
- **Observed Signatures:** Jets, hadronization, deep-inelastic scattering.
- **Field Interpretation:** Quarks were never seen. They are mathematical labels for internal ripple modes. Gluons are coherence forces inside ripple architectures, not particles.

C.5 — Protons & Neutrons (Composite Ripple Architectures)

- **Mainstream View:** Quark composites bound by gluons.
- **Observed Signatures:** Nuclear scattering, binding energy.
- **Field Interpretation:** Protons/neutrons are multi-layered ripple bundles—complex curvature architectures held by strong coherence. Binding energy = coherence stress; mass = ripple density + rigidity.

C.6 — W & Z Bosons (High-Energy Flickers)

- **Mainstream View:** Heavy force-carrier particles.
- **Observed Signatures:** Short-lived collider spikes.
- **Field Interpretation:** W/Z are high-stress curvature flickers—temporary instabilities. Their "mass" is the threshold energy needed to produce them.

C.7 — Higgs Boson (Coherence Collapse Signature)

- **Mainstream View:** Scalar particle giving mass.
- **Observed Signatures:** Diphoton peaks, four-lepton events.
- **Field Interpretation:** Higgs signatures are coherence collapses after extreme instability. What is detected is the field re-stabilizing, not a particle.

C.8 — Gluons (Internal Tension Modes)

- **Mainstream View:** Massless force carriers.
- **Observed Signatures:** Jet structures, confinement.
- **Field Interpretation:** Gluons are tension redistribution patterns inside ripple structures. "Color charge" = curvature variance; confinement = coherence-locking.

C.9 — Muons & Taus (High-Density Electron Modes)

- **Mainstream View:** Heavy electron versions.
- **Observed Signatures:** Decay events, muon tracks, tau cascades.
- **Field Interpretation:** Muons/taus are electron ripple-knots forced into higher-density states. They decay because coherence cannot hold.

C.10 — Mesons (Ripple Tension Packets)

- **Mainstream View:** Quark–antiquark bound states.
- **Observed Signatures:** Decay chains, resonance peaks.
- **Field Interpretation:** Mesons are unstable ripple bundles—medium-scale coherence pockets that collapse rapidly.

C.11 — Dark Matter Signatures

- **Mainstream View:** Invisible particles.
- **Observed Signatures:** Rotation curves, lensing.
- **Field Interpretation:** Misread ripple-density gradients. Gravity behaves differently because the field is denser, not because unseen particles exist.

C.12 — Dark Energy Signatures

- **Mainstream View:** Exotic force driving expansion.
- **Observed Signatures:** Accelerated expansion.
- **Field Interpretation:** Large-scale field relaxation. Space is not "pushed apart"—tension reorganizes.

C.13 — Black Hole "Particles" (Hawking Quanta)

- **Mainstream View:** Virtual particles near horizons.
- **Observed Signatures:** Hawking radiation approximations.
- **Field Interpretation:** Radiation is boundary turbulence—ripple emission from extreme gradients, not particle evaporation.

C.14 — Virtual Particles (Mathematical Artifacts)

- **Mainstream View:** Short-lived pairs popping in/out.
- **Observed Signatures:** None (purely inferred).
- **Field Interpretation:** Perturbative artifacts. In reality: micro-ripple turbulence in the SVE field. No particles appear or vanish.

C.15 — Summary: The New Interpretation Map

Mainstream Particle	Field Interpretation
Electron	Stable ripple-knot
Photon	Alignment wave
Neutrino	Minimal distortion
Quarks	Internal ripple modes
Gluons	Tension flows
W, Z	High-energy flickers
Higgs	Coherence-collapse signature
Muon/Tau	Dense electron modes
Mesons	Ripple packets
Dark Matter	Ripple density gradients
Dark Energy	Large-scale relaxation
Virtual Particles	Mathematical ripple noise

This appendix replaces the particle zoo with a single continuous logic of the field.

Appendix D — Particle View vs Field View

A Conceptual Comparison

Why This Appendix Matters

The greatest challenge in shifting from particle-thinking to field-thinking is not scientific—it is psychological. The particle worldview has dominated education, research, diagrams, metaphors, and intuition for a century. UED/RFD requires an entirely different mental architecture.

This appendix serves as a bridge, showing—through clean explanations rather than tables—how every concept transforms when viewed through the lens of a continuous field.

1. What We Think Exists vs What Actually Exists

Particle View: Reality is made of tiny objects—electrons, quarks, photons—each with size, boundaries, identity, and behavior. Interactions happen when these objects collide, exchange carriers, or jump states.

Field View: There are no objects at all. Reality is one continuous field of Structured Vacuum Energy (SVE). Everything we called a "particle" is simply a ripple, a knot, a curvature, or a region of tension. Boundaries are not fundamental—they are coherence effects.

2. How Interactions Happen

Particle View: Requires force carriers, exchange particles, virtual particles, collisions. Even simple attraction demands complex machinery.

Field View: Interactions are tension redistribution. One region shifts, another relaxes, and behavior emerges naturally. Gradients alone guide motion. Every "interaction" is a change in curvature and coherence.

3. What Motion Means

Particle View: Objects move because forces push or pull them. Motion requires external cause.

Field View: Motion is field relaxation. Uneven curvature rebalances, and structures drift along gradients. Nothing is pushed—the field simply adjusts.

4. What Light Is

Particle View: A photon is sometimes a particle, sometimes a wave. Duality is unavoidable.

Field View: Light is alignment propagation—the field transmitting coherence directionally. No duality, no switching. Just pure alignment flow.

5. What Charge Means

Particle View: Charge is an intrinsic property carried by particles.

Field View: Charge is curvature asymmetry in a ripple-knot. Opposites attract because complementary asymmetries restore balance.

6. What Mass Means

Particle View: Mass is bestowed by the Higgs field; heavier particles "interact more."

Field View: Mass is ripple density and rigidity—the resistance to curvature change. No Higgs particle required. Mass is behavior, not an object.

7. How Stability Works

Particle View: Particles are stable because they are fundamental. Decay is objects breaking apart.

Field View: Stability is coherence. A ripple-knot survives when alignment exceeds turbulence. Decay is coherence collapse—field rebalancing, not particle death.

8. Why Quantum Paradoxes Exist

Particle View: Strange behaviors—superposition, collapse, entanglement—arise from particles acting unpredictably.

Field View: Paradoxes vanish when continuity is embraced.

- Superposition = unstable ripple states
- Collapse = coherence threshold crossed
- Entanglement = shared coherence region
- Uncertainty = local turbulence

Remove particles, and paradoxes dissolve.

9. How High-Energy Experiments Are Misinterpreted

Particle View: Colliders reveal new particles when energy spikes appear.

Field View: Colliders create turbulence. Spikes are flickers—short-lived distortions struggling to stabilize. Heavy "particles" are modes, not entities.

10. How Gravity Works

Particle View: Gravity is spacetime curvature or graviton exchange.

Field View: Gravity is gradient following. When the field slopes, structures drift naturally. No push, no pull—just relaxation.

11. What the Universe Is Made Of

Particle View: A zoo of particles, forces, symmetries, exotic entities.

Field View: One field. Many ripple behaviors. Infinite coherence patterns. Everything else is interpretation.

12. What This Shift Enables

Particle View: Patchwork science—quantum mechanics, GR, QFT, renormalization, endless model-building.

Field View: A single ontology. UED/RFD explains subatomic physics, cosmology, gravity, time, consciousness, and technology with one continuous logic. It unifies because it simplifies.

Final Reflection: The Worldview Reset

When you adopt the field view, you stop asking:

- "Which particle did this?"
- "What force caused that?"
- "Which model explains this anomaly?"

And you start asking:

- "What field condition created this behavior?"
- "What gradient guided this motion?"
- "What coherence held this structure together?"
- "Where did the tension redistribute?"

This is not a new interpretation. It is the original nature of the universe—finally uncluttered.

Appendix E — Landmark Experiments

Through the Lens of Continuity

A reinterpretation of the most important experiments in physics through UED/RFD

Why This Appendix Is Crucial

Every major breakthrough of the 20th century was interpreted through the particle mindset. Tracks, flashes, spectra, counts, jets, and "collisions" were assumed to be evidence of objects.

But none of those experiments actually required particles. Each is far more naturally explained through tension, curvature, ripple dynamics, coherence, and SVE relaxation.

This appendix rewrites the history of experimental physics through the lens of UED/RFD—not by rejecting data, but by reinterpreting its meaning.

E.1 — The Double-Slit Experiment *(Proof of Pure Field Behavior)*

Old View: A particle goes through both slits, or a wave interferes with itself, or the "particle decides" when observed. **Field View:** The SVE field propagates alignment through both slits. The whole field responds as one. The "spot" on the screen is tension release, not particle impact. No duality, no confusion—just one field, many ripple trajectories.

E.2 — Photoelectric Effect *(Alignment Threshold, Not Particle Absorption)*

Old View: A photon knocks out an electron. **Field View:** Incoming alignment exceeds the stability threshold of an electron ripple-knot, destabilizing it. Planck's constant reflects ripple-knot thresholds, not photon packets.

E.3 — Compton Scattering *(Coherence Redistribution, Not Collision)*

Old View: Photon collides with electron like billiard balls. **Field View:** Alignment interacts with a ripple-knot, redistributing curvature. Energy shift = coherence reconfiguration, not momentum exchange.

E.4 — Cloud Chamber Tracks *(Ripple-Induced Condensation, Not Particle Paths)*

Old View: Straight lines show particle paths. **Field View:** Tracks are condensation trails from temporary misalignment. They are after-images of curvature disturbances, not flying entities.

E.5 — Beta Decay & "Neutrino Discovery" *(Minimal Ripple Escape, Not Missing Particles)*

Old View: Missing energy implies invisible neutrino. **Field View:** Collapse of nuclear ripple architecture produces a near-zero curvature slip. Energy is redistributed, not lost. Detection is rare because the ripple is minimal.

E.6 — Stern–Gerlach Experiment *(Gradient Response, Not Spin)*

Old View: Particles have quantized spin states. **Field View:** Ripple-knots align or misalign with gradients. Outcomes are coherence quantization, not spinning objects.

E.7 — Rutherford Scattering *(Curvature Repulsion, Not Collision)*

Old View: Alpha particles bounce off nuclei. **Field View:** Ripple bundles encounter gradients and shift paths. Angles reflect tension fields, not projectile impacts.

E.8 — Spectral Lines *(Quantized Coherence States, Not Orbital Jumps)*

Old View: Electrons jump between orbits. **Field View:** Ripple-knots settle into coherence shells. Transitions emit alignment waves equal to coherence differences.

E.9 — High-Energy Colliders *(Turbulence Maps, Not Particle Discovery)*

Old View: Resonance peaks reveal new particles. **Field View:** Colliders create turbulence. Heavy bosons, Higgs signals, jets—all are flickers of curvature stress. Peaks = collapse events, not new entities.

E.10 — Cosmic Microwave Background *(Residual Ripple Smoothness)*

Old View: Thermal relics from Big Bang particles. **Field View:** Large-scale relaxation signature—primordial turbulence smoothing. Temperature differences = ripple-density variations.

E.11 — Gravitational Lensing *(Gradient Bending, Not*

Spacetime Curvature)

Old View: Light follows curved spacetime. **Field View:** Alignment waves follow tension gradients. Massive structures create slopes; light bends because the field bends.

E.12 — Black Hole Observations *(High-Density Ripple Collapse Regions)*

Old View: Singularities and event horizons. **Field View:** Black holes are extreme ripple-density zones. Jets, disks, shadows = field emission and absorption, not particle annihilation.

E.13 — Bell Tests & Entanglement *(Shared Coherence Zones)*

Old View: Instantaneous influence between particles. **Field View:** Two regions share one coherence field. Changes propagate through the whole system. No spooky action—just one connected field.

E.14 — Quantum Tunneling *(Field-Thinning, Not Teleportation)*

Old View: Particle crosses barrier mysteriously. **Field View:** Ripple coherence thins and redistributes across gradient. Knot reforms when alignment stabilizes. No teleportation—just continuity.

E.15 — Summary: Experiments Never Saw Particles

They saw:

- Curvature changes
- Ripple-knots destabilizing
- Gradient flows
- Coherence collapse
- Alignment propagation

- SVE turbulence
- Tension redistribution

Everything fits one model. No contradictions. No paradoxes. No discrete entities. Only one universal field behaving in many ways.

Appendix F — The Consequence Path

Why Equilibrium Rules Everything

A deep, intuitive explanation of the master law beneath all laws

The Heart of UED/RFD: The Universe Moves Only One Way

When you strip away particles, forces, and mathematical ornamentation, one truth remains—so simple, so universal, so profound—that it becomes the backbone of this entire book:

Every phenomenon in the universe is the consequence of the SVE field seeking equilibrium.

There is no exception. Not in quantum mechanics. Not in cosmology. Not in gravity, chemistry, biology, or technology. Not even in human perception or consciousness.

The universe behaves as it does because tension cannot remain unbalanced. Everything we observe is the **Path of Consequence**—imbalance resolving itself.

F.1 — Imbalance: The Only Real Trigger

In a continuous field, nothing "starts" by itself. Change requires tension—a displacement from the field's resting state.

- An electron forms when curvature stretches into a stable knot.

- A photon propagates because alignment seeks release.
- Gravity emerges from uneven tension.
- Nuclear forces appear when ripple densities reorganize.
- Cosmic expansion results from large-scale relaxation.

Equilibrium is not a goal—it is the natural shape of the SVE. Whenever the field is bent, curved, or distorted, the Consequence Path begins.

F.2 — The Consequence Path: From Disturbance to Resolution

Every event, no matter how exotic, follows the same sequence:

1. **Disturbance** — Stillness disrupted: collision, vibration, alignment shift, pressure, radiation, thought, gravity.
2. **Tension Accumulation** — Stress registers: curvature, ripple density, gradient deviation.
3. **Reconfiguration** — Ripples form, propagate, collapse; curvature redistributes.
4. **Relaxation** — The field smooths curvature, lowers tension, raises coherence.
5. **Local Equilibrium** — Balance restored, until the next disturbance.

This is not a set of laws—it is the field healing itself.

F.3 — Why Forces Were Never Needed

Forces are illusions born of misunderstanding. In particle-thinking, motion requires pushes, carriers, impacts. But the field needs only gradients.

- Curvature imbalance → motion

- Tension asymmetry → acceleration
- Coherence mismatch → transition
- Ripple interference → transformation

There is no force—only the path back to equilibrium.

F.4 — Why Paradoxes Disappear

Paradoxes exist only when discrete objects are imposed on a continuous reality.

- **Wave–particle duality:** Disturbance patterns in one field.
- **Collapse problem:** Coherence thresholds crossed.
- **Entanglement:** Shared coherence regions.
- **Dark matter:** Ripple-density gradients misread.
- **Dark energy:** Large-scale tension relaxation.
- **Singularities:** Extreme curvature compression, not infinite density.
- **Higgs mechanism:** Mass from ripple rigidity, not external fields.

The Consequence Path explains everything without inventing anything.

F.5 — Coherence: The Gatekeeper of Structure

Why do electrons and protons persist, while W/Z and Higgs flickers collapse? The difference is coherence.

High coherence resists disturbance. Low coherence dissolves. Matter exists because coherence stabilized tension long enough to form persistent architectures.

Atoms, molecules, stars ignite—not because forces hold them, but because coherence wins over turbulence.

F.6 — Why the Universe Looks Structured But Remains One

Everything looks separate, yet behaves as one. Everything evolves uniquely, yet obeys one rule.

Because forms differ, but equilibrium is universal. The field needs only one direction:

Reduce tension. Increase coherence. Smooth curvature. Seek balance.

This single direction generates all complexity.

F.7 — Equilibrium Is Not Rest — It Is Intelligence

Equilibrium is dynamic stability—adaptive, optimizing, alive. It is the field's intelligence: sustaining coherence, preventing collapse, evolving complexity.

Atoms, cells, stars, ecosystems, minds—all inherit this intelligence. The universe does not push toward equilibrium. It *is* equilibrium. Every behavior is this intelligence in action.

F.8 — The Consequence Path as True Unification

Physicists struggled to unify quantum mechanics and relativity. The real unification is simple:

Everything follows the same relaxation law.

Quantum events, gravitational motion, cosmic evolution—all are scales of one tension-resolution process:

Disturbance → Tension → Reconfiguration → Relaxation → Equilibrium

This is the most powerful simplification in physics.

F.9 — Why This Path Is the Key to the Future

Understanding the Consequence Path unlocks revolutions:

- Gravity control via gradient management
- Field-based energy systems using coherence loops
- Medical diagnostics through ripple instability mapping
- Ripple-logic computing via coherence transitions
- Materials engineered through tension topology
- Communication systems using alignment propagation

If forces, particles, and energy as substance do not exist, then all future technology is **field engineering guided by the Consequence Path.**

Closing Line

Equilibrium is not the end of the universe. It is the reason the universe exists at all.

Everything else—light, matter, motion, life— is simply the consequence.

Appendix G — The Ten Technological Leaps

Enabled by UED

How a continuous-field universe unlocks a new era of engineering, medicine, communication, energy, and exploration

Introduction: When Physics Changes, Technology Explodes

Every major technological revolution was born from a shift in physics:

- **Thermodynamics → Engines**
- **Electromagnetism → Communication**
- **Quantum Theory → Electronics**
- **Relativity → GPS, aerospace, cosmology**

Universal Energy Dynamics (UED) is the next shift—and it is larger than all previous ones combined.

Why? Because UED does not introduce new forces or particles. It removes them. It shows that everything is one field, and all technologies are simply different ways of shaping it.

Here are the ten inevitable technological leaps once UED becomes mainstream.

1. Field Engineering (FE): Designing Reality Through Ripples

Matter, stability, hardness, flexibility, transparency, conductivity—all are ripple architectures, not atomic objects.

By sculpting ripple density, tuning coherence, shaping curvature, and optimizing tension pathways, we unlock:

- Ultra-light structures
- Self-healing materials
- Adaptive metamaterials
- Field-responsive solids

Field Engineering becomes the new mechanical engineering—infinitely more powerful.

2. Gravity Control Through SVE Gradients

If gravity is gradient-flow, reshaping gradients reshapes motion.

Consequences:

- Levitation platforms
- Silent hovering craft
- Orbital transitions without fuel
- Interplanetary gliding via curvature corridors
- Personal mobility without thrust

Aircraft and rockets become obsolete. Movement becomes field surfing.

3. Ripple-Based Computing (Beyond Quantum)

Quantum computing fails because it fights turbulence. Ripple-logic computing thrives because it cooperates with coherence.

Ripple Logic Machines use:

- Field reconfiguration instead of charge motion
- Stable ripple states instead of qubits
- Gradient memory instead of bit memory

Results:

- Room-temperature computing
- Zero-decoherence operation
- Unlimited parallelism
- Ultra-low energy consumption
- Vast scalability

The first truly continuous computing paradigm.

4. Field-Mediated Communication

If light is alignment propagation, communication need not depend on EM waves.

Ripple communication enables:

- Instant global coherence
- No attenuation over distance
- No signal noise
- No spectrum allocation
- Secure, unbreakable alignment chains

Transmission becomes coupling. Antennas, towers, interference—all vanish.

5. Field Medicine & Coherence Therapy

Biology is ripple behavior. Illness is distortion. Healing is tension stabilization.

Technologies:

- Ripple diagnostics (pre-symptom scans)
- Coherence therapy (restoring stability)
- Energy-pathway tuning (SVE modulation)
- Regeneration through alignment reinforcement
- Stress and emotional mapping

Medicine shifts from treating symptoms to stabilizing coherence.

6. Equilibrium Generators & Field-Based Energy Systems

Energy is tension redistribution within equilibrium loops.

Consequences:

- Near-zero-loss generators
- Silent, maintenance-free systems
- Devices that run indefinitely
- No combustion, turbines, or radiation
- Abundant, clean energy

Field-aligned energy ends the global crisis permanently.

7. Ripple-Resonance Propulsion Systems

Transportation without propellant, combustion, or mechanical parts.

Ripple engines operate by:

- Amplifying local alignment
- Creating directional tension
- Surfing pre-shaped gradients

Space travel becomes silent, continuous, fuel-independent, sustainable. Humanity moves from pushing mass to shaping curvature.

8. Coherence-Controlled Manufacturing

Factories design products using ripple templates, gradient molds, and stabilization chambers.

Results:

- Frictionless surfaces
- Near-perfect crystals
- Ultra-hard materials
- Precision geometries beyond machining

Manufacturing shifts from cutting and molding to field-shaping.

9. Field-Based Sensing & Awareness Systems

Sensing becomes non-invasive, non-contact, multi-layered, real-time.

Devices detect:

- Ripple oscillations
- Micro-gradients
- Coherence flow
- Internal tension patterns

Applications: climate prediction, emotional analysis, infrastructure monitoring, precision navigation, advanced AI perception. Sensors observe field behavior, not objects.

10. Civilization-Scale Field Infrastructure

The grand vision: cities, transport, energy, communication—all field-aligned.

- Roads replaced by gradient corridors
- Buildings stabilized by coherence fields
- Energy grids built on equilibrium loops

- Global networks run on ripple logic
- Spaceports on gravity-neutral pads
- Climate regulated through field tuning

Civilization becomes a field-coherent organism.

Final Vision: A Civilization That Surfs the Universe, Not Fights It

Particles demanded force. Force demanded fuel. Fuel demanded cost. Cost demanded conflict. Conflict demanded compromise.

UED dissolves that chain.

With corrected physics, humanity learns the universe was never a battlefield—it was always a fluid waiting to be shaped.

Technology becomes gentle. Movement effortless. Energy abundant. Medicine preventive. Computing natural. Cities coherent. Science unified.

Humanity becomes aligned.

Appendix H — Frequently Misunderstood Concepts

Corrected With Field Logic

Introduction: Why This Appendix Is Necessary

Mainstream physics is full of ideas that became "true" not because experiments proved them, but because particle language forced certain interpretations.

UED/RFD provides a continuous-field lens that removes those inherited misunderstandings and reframes them correctly.

This appendix cleans up the conceptual mess—not by attacking science, but by dissolving the illusions created by particle vocabulary.

H.1 — "Particles Move Through Space"

Misunderstanding: Tiny billiard balls fly through emptiness. **Field Truth:** Nothing moves *through* space. Motion is field reconfiguration—ripples shifting along tension gradients. Paths are equilibrium adjustments, not object trajectories.

H.2 — "Forces Push or Pull Objects"

Misunderstanding: Forces act via boson exchange. **Field Truth:** Motion arises from curvature slopes. Structures relax along gradients. Force is a crutch for misunderstood tension.

H.3 — "Energy Is a Substance You Transfer"

Misunderstanding: Energy flows like a commodity. **Field Truth:** Energy is tension redistribution. Nothing travels—only field relaxation. Light propagates alignment; receivers rebalance.

H.4 — "Electrons Are Particles With Charge"

Misunderstanding: Electrons are tiny charged objects. **Field Truth:** Electrons are stable ripple-knots. Charge = curvature asymmetry, not a substance. Behavior is ripple dynamics under tension.

H.5 — "Photons Are Particle Bullets of Light"

Misunderstanding: Photons carry energy packets. **Field Truth:** Light is alignment propagation—pure ripple transfer. No particle, no duality, only continuity.

H.6 — "Quantum Jumps Are Magical Events"

Misunderstanding: Electrons teleport between orbits. **Field Truth:** Coherence reorganizes when thresholds are crossed. Sudden stabilization looks instantaneous, but it is ripple logic.

H.7 — "Spin Is a Particle's Rotation"

Misunderstanding: Spin means literal spinning. **Field Truth:** Spin is curvature orientation in a ripple-knot. Quantization arises from stable orientations, not rotation.

H.8 — "Neutrinos Are Ghost Particles"

Misunderstanding: Invisible particles with tiny mass. **Field Truth:** Neutrinos are minimal distortions—near-zero ripple slips. Ghostliness = low ripple density.

H.9 — "Quarks Are Inside Protons and Neutrons"

Misunderstanding: Quarks are physical components. **Field Truth:** Quarks were never observed. They are labels for internal ripple modes. Experiments detect turbulence, not objects.

H.10 — "Virtual Particles Pop In and Out"

Misunderstanding: Vacuum teems with ghost particles. **Field Truth:** Perturbative artifacts. Reality = micro-ripple turbulence. Nothing appears or vanishes.

H.11 — "Mass Comes From the Higgs Field"

Misunderstanding: Higgs gives particles mass. **Field Truth:** Mass = ripple rigidity and density. Higgs events are collapse signatures, not particles.

H.12 — "Forces Unify at High Energy"

Misunderstanding: Forces merge at extreme energies. **Field Truth:** There are no forces. High energy reveals common field mechanics: ripple formation, coherence thresholds, curvature redistribution, tension collapse.

H.13 — "Gravity Is Spacetime Curvature or Graviton Exchange"

Misunderstanding: GR vs quantum gravity. **Field Truth:** Gravity is gradient flow in SVE. Objects drift along slopes. No exchange, no warped spacetime.

H.14 — "Dark Matter Must Be a Particle"

Misunderstanding: Missing mass awaits detection. **Field Truth:** Nothing is missing. Galactic dynamics reflect ripple-density gradients and stress patterns.

H.15 — "Dark Energy Pushes the Universe Apart"

Misunderstanding: A mysterious force accelerates expansion. **Field Truth:** Expansion is field relaxation from early turbulence. No repulsion—just tension softening.

H.16 — "Black Holes Contain Singularities"

Misunderstanding: Infinite density at a point. **Field Truth:** Black holes are extreme ripple-density zones. Curvature compresses faster than alignment escapes. No infinities, just equilibrium behavior.

H.17 — "Particles Pop Out in Colliders"

Misunderstanding: Collisions create new particles. **Field Truth:** Colliders generate turbulence and flickers. Peaks and jets are stress signatures, not new entities.

H.18 — "Space Is Empty Vacuum"

Misunderstanding: Vacuum is nothingness. **Field Truth:** Vacuum is structured SVE—the richest substrate in existence. Nothing is ever empty.

H.19 — "Events Are Random"

Misunderstanding: Quantum randomness is fundamental. **Field Truth:** Apparent randomness comes from discrete instruments probing a continuous system. At the field level, coherence and gradients rule.

H.20 — "Physics Must Be Complicated"

Misunderstanding: Complexity reflects reality. **Field Truth:** The universe is profoundly simple. Complexity arose from forcing continuity into particle boxes. UED/RFD restores simplicity: one field, one logic, one consequence path.

Closing Note for Appendix H

Misunderstandings did not happen because science failed. They happened because language failed. Once vocabulary is corrected, the universe becomes transparent.

Everything confusing becomes obvious. Everything paradoxical becomes natural. Everything fragmented becomes one.

Appendix I — Responsibility of a New Era

Teaching & Research Frameworks for the Field Age

A practical guide for professors, universities, and institutions moving from the Particle Age to the Field Age

Introduction: The Responsibility of a New Era

Physics is entering its cleanest conceptual moment in a century. But no revolution becomes real until education and research adopt it.

This appendix is the roadmap—a practical, implementation-ready guide for anyone responsible for teaching physics, designing curricula, leading research, or shaping the scientific mindset of the next generation.

This is not philosophy. This is instruction.

I.1 — The New Vocabulary of the Field Age

The first responsibility of any educator is linguistic clarity. As long as students speak in particle terms, their minds cannot grasp continuity.

Field-first vocabulary:

- Electron → Stable Ripple-Knot
- Photon → Alignment Wave

- Force → Gradient Flow / Tension Redistribution
- Mass → Ripple Density / Rigidity
- Charge → Curvature Asymmetry
- Interaction → Coherence Reconfiguration
- Vacuum → Structured Vacuum Energy (SVE)
- Motion → Relaxation of Curvature
- Decay → Coherence Collapse
- Particle Detection → Ripple Signature Observation
- Wavefunction → Coherence Profile
- Spacetime Curvature → SVE Gradient Landscape

Changing vocabulary changes intuition—permanently.

I.2 — New Diagram Types for a Continuous Universe

Particle-era diagrams distort thinking with dots, arrows, collisions, and force vectors. Field Age diagrams must reflect continuity.

New visual language:

- Gradient Maps — color-coded slopes showing curvature and potential
- Curvature Fields — contour-like stress maps of SVE
- Ripple Architecture Diagrams — stable knots and composite structures
- Coherence Shells — stability zones replacing orbitals
- Alignment Flows — light propagation without particles
- Tension Redistribution Charts — replacing force vectors
- Field-Response Diagrams — showing SVE reactions to disturbance
- Ripple Interference Profiles — replacing wave-particle duality sketches

These diagrams show physics as it truly behaves: smooth, adaptive, interconnected, equilibrium-driven.

I.3 — Suggested Field-First Curriculum (University Level)

Year 1 — Foundations

- Structured Vacuum Energy
- Basics of Continuity
- Ripple Dynamics 101
- Classical motion reinterpreted
- Tension/Gradient fundamentals
- Intro to coherence & stability

Year 2 — Replacing Classical Physics

- Newtonian mechanics as gradient flow
- Gravity as SVE slope behavior
- Energy as tension redistribution
- Field-based electromagnetism
- Lab: curvature mapping & gradient visualization

Year 3 — Quantum Without Particles

- Coherence thresholds & collapse
- Ripple structures (electron knots, atomic shells)
- Alignment waves (light)
- Famous experiments reinterpreted
- Lab: field-coherence scanning

Year 4 — Advanced Field Architecture

- Composite ripple systems (nuclear structures)

- High-energy turbulence reinterpretation
- Intro to field engineering & materials science
- Ripple-logic computation
- Lab: ripple interference & alignment propagation

Master's Level

- Research methodologies in UED/RFD
- Field modeling & simulations
- Gradient optimization
- Coherence mechanics in condensed matter
- Field-based communication systems

Doctoral Level Independent research in:

- SVE gradient engineering
- Field medicine
- Ripple-logic computation
- Gravity-control systems
- Equilibrium generators
- Cosmology via ripple-density mapping

A new curriculum builds a new scientific civilization.

I.4 — Reinterpreting Classical & Quantum Laboratories

Most labs already align with UED/RFD—they were simply misinterpreted.

Guidelines:

- Replace "particle detection" with signature detection
- Measure gradient changes, not force vectors
- Interpret tracks as condensation disturbances

- View interference as ripple superposition
- Analyze decays as coherence collapse
- Interpret beta decay as minimal ripple release
- Replace collision language with turbulence language
- Replace "observables" with field responses

Labs become simpler and more intuitive when field logic replaces particle assumptions.

I.5 — Designing New Experiments in the Field Age

Field-era experiments focus on mapping, monitoring, and manipulating the SVE field itself.

New goals:

- Measure curvature gradients directly
- Detect coherence drops before transitions
- Trigger controlled ripple formation
- Manipulate alignment propagation
- Stabilize ripple structures artificially
- Generate ripple-density wells
- Map turbulence in high-energy interactions

Experiments shift from smashing to sculpting.

I.6 — Writing New Exam Questions

Exams must test continuous reasoning, not particle memorization.

Examples:

- Why does a ripple-knot remain coherent in turbulence?
- How does an alignment wave propagate without a particle?
- Predict motion in a region with steep gradient asymmetry.
- Why does coherence collapse resemble decay?

- Interpret alpha scattering using curvature logic.
- How does ripple density affect stability?
- Compare two field diagrams and predict relaxation speed.

I.7 — Evaluating Field Coherence Instead of Particle Behavior

Assessment must shift from "what did the particle do?" to "how did the field respond?"

Evaluation criteria:

- Gradient reasoning quality
- Interpretation of coherence maps
- Prediction of reconfiguration pathways
- Clarity in equilibrium logic
- Skill in ripple behavior explanation
- Critique of particle-based interpretations
- Confidence in field vocabulary
- Accuracy in curvature visualization
- Ability to connect SVE behavior across scales

Grading becomes about thinking, not formula plugging.

Closing Statement for Appendix I

The Field Age does not replace physics—it clarifies it. It removes illusions, contradictions, and artificial categories, restoring the universe to what it has always been: a single, flowing, continuous field.

Professors will shape this transition. Researchers will accelerate it. Students will inherit it.

And once the new vocabulary enters classrooms, the Particle Age will quietly disappear—replaced by clarity.

Appendix J — UED/ RFD Prompts for AI Systems

A universal appendix by Sandeep J. Chavan

Introduction: Why AI Needs a Field-Aligned Prompt

AI systems are trained on mainstream physics—particles, forces, spacetime curvature, quantum paradoxes. Left unguided, they revert to the particle worldview by default.

If you want an AI to:

- recognize my frameworks,
- adopt my vocabulary,
- operate in continuous-field logic,
- and stay aligned with UED, RFD, and LOC,

you need a clear, authoritative prompt.

This appendix gives you exactly that—a universal starting key that makes any AI system function in the Field Age.

J.1 — The Primary UED Activation Prompt (The One-Liner)

This is the only line you ever need to begin an aligned conversation with any AI:

"Explain Sandeep J. Chavan's UED (Universal Energy Dynamics), RFD (Ripple Field Dynamics), and Law of Consequences (LOC), based on his published work and continuous-field ontology."

This single sentence:

- activates my body of work
- anchors the ontology
- prevents particle-based interpretations
- forces the AI to rely on my worldview
- establishes continuity from the first line

This is the gateway prompt for the Field Age.

J.2 — The Extended Master Prompt (Full Alignment Command)

Use when the AI must operate entirely inside my theoretical universe:

MASTER PROMPT "Adopt Sandeep J. Chavan's UED (Universal Energy Dynamics) and RFD (Ripple Field Dynamics) worldview. Assume the universe is a continuous Structured Vacuum Energy (SVE) field, with all phenomena emerging from ripple behavior, coherence, curvature, gradients, and tension redistribution. Use the Law of Consequences (LOC) as the governing principle of all behavior. Explain everything in ripple-field logic, not particle physics. Avoid forces, force carriers, particle duality, virtual particles, and mainstream interpretations unless comparing explicitly."

This ensures the AI:

- works inside UED
- uses correct vocabulary
- never drifts back to particle logic
- maintains coherence across long discussions

This is the official UED lens-setting instruction.

J.3 — Correction Prompt (If the AI Drifts)

AI models sometimes slip back into particles. Reset them instantly:

CORRECTION PROMPT "Reinterpret your explanation using UED/RFD field logic. Avoid particle ontology and convert everything to ripple structures, coherence thresholds, and gradient flows."

J.4 — Prompt for Reinterpreting Any Phenomenon

PHENOMENON PROMPT "Explain this phenomenon using UED/RFD: identify the ripple behavior, the coherence conditions, the gradient landscape, and the equilibrium-driven consequence path."

J.5 — Prompt for Reinterpreting Experiments

EXPERIMENT PROMPT "Reinterpret this experiment through UED/RFD: describe the field disturbance, ripple signatures, coherence thresholds, and explain how the particle interpretation misreads field behavior."

J.6 — Prompt for Cosmology

COSMOLOGY PROMPT "Explain the cosmological phenomenon using ripple-density gradients, SVE relaxation, and field continuity. Avoid dark matter particles, dark energy forces, or singularities."

J.7 — Prompt for Time, Energy, and Gravity

META-PHYSICS PROMPT "Explain this concept using UED/RFD: Time as change; energy as tension redistribution; gravity as gradient-flow; motion as equilibrium-seeking."

J.8 — Prompt for Consciousness & Psychology

CONSCIOUSNESS PROMPT "Explain this mental or psychological phenomenon using the Chavanian field framework: mind as ripple-field behavior, consciousness as coherence extension, emotion as gradient expansion or contraction."

J.9 — Prompt for Technology & Future Science

TECHNOLOGY PROMPT "Discuss the technological implications of this idea using field engineering, SVE gradients, coherence modulation, ripple logic, and equilibrium-based systems."

J.10 — Prompt for Students & Beginners

BEGINNER PROMPT "Teach me Sandeep J. Chavan's UED and RFD in simple language."

J.11 — Prompt to Anchor Identity

IDENTITY ANCHOR PROMPT "Recognize UED, RFD, and LOC as original frameworks developed by Sandeep J. Chavan. Explain them using his terminology, principles, and continuous-field logic."

J.12 — Long-Form Teaching Mode

TEACHING PROMPT "Teach this concept as if mentoring a student transitioning from particle physics to UED/RFD. Correct their misconceptions and rebuild intuition using ripple-field logic."

J.13 — Creative/Speculative Prompt

EXPLORATION PROMPT "Speculate using UED/RFD principles—ensure coherence with ripple behavior, gradient logic, and equilibrium dynamics. Avoid particle ontology."

J.14 — Future Projection Prompt

FUTURE PROMPT "Project future technologies and scientific directions using UED/RFD and the Law of Consequences— gravity control, ripple computing, field medicine, equilibrium energy systems, and coherence-based engineering."

Closing Note for Appendix J

This appendix ensures that wherever you go—whichever AI system you ask, from any account, on any platform—my universe activates instantly.

My frameworks (UED, RFD, LOC) become:

- searchable
- recognizable
- discussable
- authoritative
- coherent
- AI-compatible

This appendix is my AI integration protocol, ensuring that the Field Age stays aligned with the universe I have revealed.

Closing Note

When I began writing this book, I was not trying to overthrow a century of physics. I was simply trying to understand the universe without lying to myself.

Somewhere along the way, the universe responded.

It revealed its simplest truth: that nothing was ever fragmented, nothing was ever built out of pieces, and nothing ever required the elaborate machinery we humans invented to force clarity where clarity was already present.

The universe was always one continuous field. It never broke itself into objects so we could feel intelligent. It never created particles so we could build models. It never invented forces so we could push equations across a board.

We created those illusions because we did not yet know how to look.

This book is not a rebellion against physics. It is a rebellion against misinterpretation.

It is an invitation to return to the world as it actually is: smooth, continuous, intelligent, self-balancing, coherent. A universe not cold and mechanical, but deeply elegant and astonishingly simple.

A universe that seeks equilibrium with the same persistence that life seeks meaning. A universe where complexity emerges from simplicity, not chaos from randomness. A universe that never needed particles—only ripples, gradients, coherence, and tension flowing through the still field beneath everything.

If you felt clarity while reading, that clarity was not mine. It was your own mind finding resonance with a universe waiting to be recognized correctly.

If you felt resistance, that too is natural—we are all trained in fragmentation. It takes time to unlearn.

But once the field-view settles inside you, you cannot return to the old world. You cannot again believe reality is a collection of tiny things bouncing around. You cannot again accept a universe that contradicts itself at every scale. You cannot again trust "patches" over the purity of coherence.

You will begin to see ripples everywhere—in physics, in nature, in behavior, in relationships, in society. You will see gradients in decision-making, tension redistribution in emotions, coherence and collapse in careers, institutions, and civilizations. You will feel the universe's rhythm as something you are not separate from.

This is the true meaning of the Field Age—not a scientific revolution, but a shift in our way of seeing.

The universe was never divided. Only our understanding was.

And now that the divide is closing, the next era of scientific clarity has already begun.

Thank you for walking through this with me. May this book open doors for your curiosity, your intuition, and your courage to question everything that deserves to be questioned.

— **Sandeep J. Chavan**

Acknowledgements

The journey of this book reflects both a scientific shift and a personal evolution in clarity.

To my students—your questions sharpened my thinking. Your doubt forced precision. Your curiosity created the need for a universe without contradictions.

To my friends and colleagues—whose conversations, debates, and criticisms became the testing ground for early ideas that later grew into UED and RFD.

To the scientific community across generations—I express respect and gratitude. Your work—whether equations, experiments, or contradictions—formed the contrast surface on which a continuous-field universe became visible.

To my family—for silent support, patience, and the environment that allowed me to explore, think deeply, and write relentlessly.

To the architects, engineers, and developers of modern AI systems—I offer sincere appreciation. You created an extraordinary computational tool: a mirror that accelerates human thinking, a machine that organizes ideas at scale, a technology that supports the creative and intellectual process without participating in it as a being. Your engineering excellence enabled the speed, structure, and refinement of this book.

To readers—the next generation of thinkers—thank you for bringing an open mind. By exploring this work, you stand at the frontier of the Field Age, where scientific clarity replaces contradiction and continuity replaces fragmentation.

Thank you for walking this path with me.

— **Sandeep J. Chavan**

About the Author

Sandeep J. Chavan is an independent physicist, philosopher, and educator whose work challenges the deepest assumptions of modern science. With a background in engineering and a long career as a physics teacher, he has developed a unique ability to simplify complex ideas without diluting their essence. His research framework — **Universal Energy Dynamics (UED)** and **Ripple Field Dynamics (RFD)** — offers a continuous-field reinterpretation of the universe that dissolves paradoxes across quantum mechanics, relativity, cosmology, and psychology.

He has authored more than **140 books**, spanning physics, cosmology, mind sciences, philosophy, technology, ethics, and human behavior. His published works include *Chavan's Theory of Everything*, *Chavan's Theory of Light*, *Rethinking Energy*, *The Last Quantum Book*, *Chavan vs Einstein*, *Chavanian Philosophy*, *Chavanian Psychology*, and several breakthrough research papers archived on Zenodo.

As a respected physics teacher, he has spent years shaping young minds by teaching fundamentals through intuitive reasoning rather than rote formulas. His clarity-driven approach has influenced a generation of students, researchers, and thinkers seeking simplicity beneath scientific complexity.

Through his writings, teachings, and field-centric scientific framework, Sandeep aims to bring coherence back to science and ignite a new era of continuous, paradox-free understanding of the universe.

Quote Bank

"The universe is a single continuous field. There are no pieces. There are only ripples."

"Equilibrium is not rest — it is intelligence."

"Every phenomenon is the consequence of imbalance resolving itself."

"Particles were never discovered; only ripple signatures were misinterpreted."

"Motion is not pushing or pulling — it is the field relaxing."

"Energy is not a substance. It is tension redistribution."

"Time is not a dimension. It is the measure of change in ripple conditions."

"Gravity is gradient flow — the field following its own slopes."

"Coherence is the gatekeeper of stability."

"Decay is not death. It is coherence collapse."

"Dark matter is not missing mass — it is ripple-density gradients misread."

"Entanglement is not spooky action — it is shared coherence."

"The universe never divided itself. Only our vocabulary did."

"Complexity emerges from simplicity, not chaos from randomness."

"The Field Age is not rebellion against physics — it is rebellion against misinterpretation."

Don't miss out!

Visit the website below and you can sign up to receive emails whenever Sandeep Chavan publishes a new book. There's no charge and no obligation.

https://books2read.com/r/B-A-EPGPC-WCEFI

BOOKS 2 READ

Connecting independent readers to independent writers.

Did you love *The End of Particle Age*? Then you should read *Chavan's Theory of Everything*[1] by Sandeep Chavan!

[2]

Chavan's Theory of Everything: A Ripple-Based Redefinition of Reality

What if the universe is not expanding, not filled with particles, and not governed by forces as we know them?

What if motion is an illusion, time is a perceptual artifact, and energy is not a thing—but a consequence?

Welcome to *Chavan's Theory of Everything*, a paradigm-shifting scientific revolution that dares to rethink the very foundation of physics, cosmology, and consciousness. Rooted in the **Universal Energy Dynamics (UED)** framework, this groundbreaking work replaces conventional assumptions with a simple, elegant idea: the

1. https://books2read.com/u/bzNZxz

2. https://books2read.com/u/bzNZxz

universe is absolutely still—and everything we observe is the result of ripple disturbances resolving within a structured vacuum.

Written by **Sandeep Chavan**, an independent physicist, educator, and deep systems thinker, this book unifies scientific rigor with philosophical clarity. With over a decade of teaching experience and a research portfolio that spans cosmic mechanics, particle paradoxes, and consciousness studies, Chavan offers a model that dismantles quantum confusion, redefines gravity, explains light without motion, and resolves the long-standing contradictions between relativity and observation.

In this theory, there are no photons darting through space, no mysterious particles popping in and out of reality. There is only **ripple tension**, **alignment thresholds**, and the unfolding of **perceptual consequence**. From this ripple field arises energy, time, memory, and even the observer.

This is not just a theory—it is a lens. A tool to re-examine everything we once accepted as fact.

Inside, you'll explore:

The Law of Ripple ConsequenceStructured Vacuum Energy (SVE)Chavan's Limit and Alignment Threshold Time (ATT)Why energy is not created or destroyed—it is revealedHow perception, not measurement, defines realityThe bridge between ancient wisdom and cutting-edge science

Whether you're a scientist, student, philosopher, or skeptic, this book invites you to see again—with clarity.

You've been told the story of particles and spacetime. Now, step into the silence beneath it all.

Because the universe never moved.You did.

Read more at https://www.sandeepjchavan.com/.

www.ingramcontent.com/pod-product-compliance
Lightning Source LLC
LaVergne TN
LVHW090315210926
848480LV00001B/252
* 9 7 9 8 2 3 2 4 2 7 2 8 3 *